Clinical Integration of Complementary Mind-Body Therapies:

On Psychological and Behavioral Health

Raqota Berger

Copyright © 2016 by Raqota Berger

All Right Reserved.

Printed in the United States of America

First Edition

Published by Rizin
P.O. Box 3177
San Diego, CA.
93003

Except as permitted under the United States Copyright Act of 1976, no part of this publication may be reproduced or distributed in any form or by any means, including, but not limited to, electronic, mechanical, photocopying, recording, scanning, digitization, or without the prior written permission of the publisher. Exceptions are made for quotations and referenced paraphrasing embodied in critical articles, books, and reviews.

Library of Congress Cataloging-in-Publication Data

ISBN: 978-0-578-17614-7

Paperback

This book is printed on acid-free paper.

To a Natural and Healthy Life

Table of Contents

Chapter One

Introduction to Complementary Mind-Body Therapies

This auspicious endeavor investigates and makes recommendations to the psychological community of the utility of incorporating certain complementary therapies into clinical practice. This analytic investigation contains a breadth of apposite research that supports this contemporary therapeutic recommendation. It is well known that talk therapy is an effective form of psychological treatment and provides many benefits to clients across a wide range of psychological and behavioral issues (e.g., Casement, 1991; Martin, 2011; Plante, 2011; Spiegler & Guevremont, 2010). It is not as well understood how mind-body therapies can also be effective in treating a range of mental health issues and behavioral problems. Cognitive–behavioral therapy and mindfulness-based interventions are the two therapeutic approaches that have been studied far more often with CAM modalities than any other approaches in the field (Allison & Faith, 1996; Barnett, Shale, Elkins, & Fisher, 2014; Lau & McCain, 2005). As this is the case these clinical approaches to psychological treatment are the core therapies examined throughout this work.

It is not being positioned here that complementary mind-body therapies are optimal, or even desirable, for all psychological or behavioral conditions. This work has its sights set on those symptoms and conditions that have been demonstrated to be clinically improved through the use of the specific CAM modalities covered in this analytic review. The overarching goal of this in-depth tractate is to show how combining certain forms of talk therapy with specific mind-body techniques can often help to better serve clients.

Overview of Topic

There is a flowering trend taking place across the nation in regard to clinical acceptance and use of mind-body complementary therapies. The most current research on this topic demonstrates that the range of complementary therapies commonly used in psychological practice today have been increasingly incorporated into traditional psychological practice and have been shown to have significant effects on specific mental and physical health conditions (e.g., Edenfield & Saeed, 2012; Field, 2012; Kaley-Isley, Peterson, Fischer, & Peterson, 2010). As an example of this growing popularity, Pratt (2013) put forth the argument that the use of complementary therapies among the general public is both substantial and continues to become more accepted. To demonstrate this continuing shift among the population she conducted a study with 592 participants ranging in age from 18 to 52 and it was found that 100% of those in her study reported the use of at least one type of complementary and alternative medicine (CAM) practice within their lifetime. It was also found that 88% of those in her study used at least one type of CAM practice within the last year. If the findings from this study typify the attitudes and beliefs of the general population then we can only expect the use of CAM therapies to continue to increase over time.

The evidence for the effectiveness of complementary therapies is growing each year. As an example, Mehta and Sharma (2010) demonstrated in meta-analysis of 18 randomized controlled trials that yoga is consistently able to significantly reduce depressive symptoms in patients. These clinically significant findings held with patients across 17 of the 18 studies that met the criteria for the study. Field (2012) conducted a meta-analysis on the effects of complementary therapies on children and adolescents and concluded that the evidence is overwhelmingly clear that they are clinically relevant. Her meta-analytic findings showed that several mind-body therapies (e.g., tai chi and yoga) had clinically significant effects on a wide range of issues important to the interests of psychological practice, including depression, attention-deficit hyperactivity disorder (ADHD), anxiety, cognitive performance, eating disorders, attention disorders, stress, sleeping disorders, etc. This utility of CAM therapies has been demonstrated across a wide range of research and the evidence for this is continuing to build (Freeman, 2009).

Those employing complementary therapies tend to use them together with conventional therapies. CAM treatments are not typically thought of as a replacement for traditional treatments but rather as complements to conventional care and practice. The use of CAM approaches in treatment is increasing in the West (Arias, Steinberg, Banga, & Trestman, 2006). According to these authors some of the commonly used CAM modalities include meditation, deep breathing exercises, massage, yoga, and chiropractics. Raschetti, Menniti-Ippolito, Forcella, and Bianchi (2005) highlight some of the most popular mind-body therapies in use as yoga, progressive muscle relaxation, guided imagery, martial arts, meditation, and deep breathing exercises. In her work on mind-body CAM interventions, Pratt (2013) informs us that each year in the United States around $35 billion is spent out-of-pocket on complementary treatments. Nahin, Barnes, Stussman, and Bloom (2009) stress how important it is to learn more about the importance of CAM therapies in mental health treatment. The importance of this is not only because of how effective they are in treating multiple mental and physical health problems but also because this adjunct market makes up over 11% of total out-of-pocket health care spending each year in the United States. The research team went on to point out that in 2007 over 38 million adults in the United States alone made an estimated 354 million visits to CAM practitioners (Nahin, et al., 2009). Based on the projections and current trends it would seem clear that complementary therapies are only going to become more popular and accepted in the coming years.

Edenfield and Saeed (2012) supported this position in their work by highlighting how certain mind-body therapies can alleviate a range of mental health issues. In their work they put forth the merits of meditation, relaxation techniques, and mindfulness-based interventions to treat a range of mental health problems involving depression, generalized anxiety, chronic stress, eating disorders, and so forth. The authors concluded that the most clinically relevant and recent data show that these complementary psychological therapies are both psychologically and physically beneficial and should be incorporated into clinical practice. Meditation and relaxation training has been shown to be clinically useful across a number of studies looking at range of psychological health issues. Many independent researchers have shown that these highlighted complementary therapies are clinically useful and highly recommended for treating a wide range of mental and physical health problems (e.g., Fouladbakhsh & Stommel, 2010; Graham, 2012; Harris, Alexander, Cronkite, & Moos, 2006). The purpose of this work is to comprehensively demonstrate the utility of incorporating these healthy and natural complementary forms of therapy into clinical practice.

Chapter 2

Overview of the Core Complementary Therapies

Current Empirical Evidence

There is a wide range of complementary approaches that have been shown to be useful in both clinical treatment for psychological problems and in medical treatment for physical problems. In this work the focus is on mind-body approaches to healing. Put simply, mind-body work views effective healing and treatment as requiring equal emphasis on both the mental and the physical. That is, there is belief that there is an intricate and inseparable physiologic pathway between the mind and the body and each must be addressed for the most optimal treatment (Abrams, 2001; Cuddy & Duffin, 2005). Guthrie (1995) addresses this by arguing that in many cases healing must come through the body. She put forth that when the physical body is damaged in some way in (e.g., physical abuse, rape, etc.) that the client cannot be adequately healed on a purely intellectual level, or even adequately healed through traditional talking therapies. To press her position on this even further she asserted, "I am now convinced that ultimately healing has to come through the body; I think this is a shortcoming of traditional talking therapy" (p.6). The remainder of this section brings forth a number of commonly used mind-body therapies that have been demonstrated to be effective in psychological treatment.

Body Movement Exercises

Mind-body therapies involve some type of physical movement. The most basic form of physical movement is exercise. Body movement has been shown to be effective in treating a wide range of mental and physical health problems. According to Freeman (2009) exercise as a form of therapy treats four primary conditions involving mood disorders, disabilities of aging, cardiovascular disease, and pulmonary dysfunction. Field (2009) put forth that exercise as a form of therapy has been shown effective for wide range of behavioral, mental and physical health problems, including depression, cancer, stress, smoking cessation, obesity, weight reduction, osteoporosis, fibromyalgia, and diabetes. Numerous studies have shown the beneficial effects of body movement exercises on mood and mood disorders. Arent, Landers, and Etnier (2000) demonstrated in a meta-analytic review that looked at 158 effect sizes from 32 independent studies that chronic exercise is associated with improved mood in the elderly. These statistical findings held across all 32 studies and all effects sizes were significantly greater in the experimental groups than they were for any of the control groups.

Harris, Alexander, Cronkite, and Moos (2006) conducted an epidemiological study that examined the relationship between exercise and depression. This study has a sample size of 424 clinically depressed adults and concluded that exercise coping and body movement was associated with less concurrent depression. These findings held that physical activity counteracts the effects of negative life events, major life stressors, depression, and medical problems. Penninx, et al. (2002) found similar results in their study on exercise and depression. The results from the clinical trial with 439 elderly individuals aged 60 and over showed that aerobic and resistance exercise significantly reduced not only depression across the sample, but also significantly reduced disability and pain. Other studies have also put forth the evidence that body exercises are clinically useful in the treatment of depression and anxiety (e.g., Field, 2012; Martinsen, 2008).

In his work looking at psychological treatment utilizing cognitive-behavioral therapy (CBT), Martinsen (2008) demonstrated that mild and moderate forms of both depression and anxiety can often be more effectively treated with the supplementation of exercise therapy. Field (2012) put forth similar findings and explained in her work that exercise therapy greatly reduces depressive symptoms and anxiety in children and adolescents. She strengthened her position on the use of exercise as a complementary therapy by pointing out that this does not come with the undesirable side effects that have been associated with the common treatments of depression and anxiety (e.g., antianxiolytic medications and antidepressants). Freeman (2009) provides a very detailed examination of the scope of the positive outcomes that accompany exercise therapy, which includes mood state modulation, improvements in cognitive functioning, and positive effects on depression, stress, and anxiety. Her findings put forth also demonstrated the positive impact on lesser known and discussed psychological problems, including seasonal affective disorder, self-efficacy, and neuroticism. The evidence in the literature seems very clear that body movement therapies are a beneficial supplement to traditional psychological treatment. The therapeutic merits of mind-body therapies are one of the central arguments stressed throughout this work.

Meditation

Meditation is one of the most commonly researched complementary therapies that can be found in the relevant literature. Mindfulness meditation is the specific form of meditation that has been researched the most for psychological conditions and medical illnesses (Kabat-Zinn, 2003). According to Park (2013) meditation was one of the first mind-body therapies to be recognized by mainstream healthcare providers in the United States and its efficacy has been demonstrated across a range of conditions (e. g., hypertension, panic disorders, and alleviation of menopause symptoms). A meta-analysis conducted on 21 studies investigating the benefits of mindfulness meditation for certain chronic conditions (e.g., pain), anxiety, and depression concluded that post-treatment effect sizes were significant and clinically relevant (Cohen's *d* = 0.74, SD = 0.39; Baer, 2003). The researcher was clear in his belief that the findings of the meta-analysis demonstrated that mindfulness-based meditation can bring those that are mildly to moderately distressed within (or close to) the normal range.

Further studies show the efficacy of meditation therapy in psychological practice. Krisanaprakornkit, Ngamjarus, Witoonchart, and Piyavhatkul (2010) argued that meditation is a useful tool for treating attention-deficit hyperactivity disorder (ADHD). They proposed that mediation could be included in the treatment of this psychological condition for attentional training and their results concluded that mediation showed to be just as effective as both drug therapy and standard therapy across several ADHD rating scales with a total of 49 child and adult patients. Edenfield and Saeed (2012) showed in their work how mindfulness meditation techniques can significantly impact the autonomic nervous system and help to reduce stress, anxiety, and depression in those suffering from these particular chronic psychological conditions.

Field (2009) explored several conditions that she found to be fit for meditation therapy. She penned that meditative practices have been shown to have potency in the treatment in healing of depression, stress, anxiety, irritable bowel syndrome, cardiac conditions, and fibromyalgia. There was also clear evidence put forth in her work that meditation also increases vagal activity and theta wave activity which results in a more positive and happy mood state (Field, 2009). There is also evidence that mediation therapy can improve immune conditions through improving overall feelings of well-being, enhanced coping, reduction in chronic stress, and improved psychological functioning (Otto, Norris, & Bauer-Wu, 2006). This team drew these conclusions after reviewing 9 studies examining the use of mindfulness meditation across a range of cancer patients (e.g., breast and prostate). Krisanaprakornit, Krisanaprakornit,

Piyavhatkul, and Laopaiboon (2006) put forth their findings about the benefits of meditation therapy in the treatment for anxiety disorders across two randomized controlled trials. They found in the different studies that mediation therapy demonstrated comparable clinical significance to anti-anxiety drugs and yoga, and scores on the Yale-Brown Obsessive Compulsive Scale did not produce any statistically significant differences between groups. The researchers also made it clear that more trials are needed due to high dropout rates and that any statements about the effectiveness of meditation in the psychological treatment of anxiety disorders are not conclusive (Krisanaprakornit, 2006). Based on the literature it seems clear that mindfulness meditation has its place and use in treating certain psychological and medical conditions.

Progressive Muscle Relaxation

Another widely covered complementary therapy in the literature is progressive muscle relaxation (PMR). With this approach the client is trained to voluntarily calm the body's muscles and to relax them at command. Freeman's (2009) collection of the prevailing research on relaxation therapy demonstrated a wide range of therapeutic benefits across both psychological and physical conditions. The strongest evidence seemed to reside in the areas treating anxiety, cognitive functioning, mood state management, irritable bowel syndrome, immunity, chronic fatigue, pain management, and hypertension. Park (2013) argued that learning to deactivate the muscular system through PMR (a system she described as the initial tensing of certain muscle groups and then the subsequent releasing of that tension) is useful to patients trying to obtaining mental calmness and reduce psychological distress.

In a randomized controlled study by Ghoncheh and Smith (2004) it was found that PMR resulted in higher levels of overall relaxation states (both mental and physical) and higher levels of mental quiet (calm) based on scores from the Smith Relaxation States Inventory that has 30 self-report items that measure the three stress states (somatic, stress, worry, and negative emotion). The authors claimed that PMR is the "preferred approach of clinical psychologists" (p. 131) in modern treatment because of its clear impact on cognitive and somatic conditions (Ghoncheh & Smith, 2004). PMR has become a common feature of clinical training and its outcomes have been shown to be comparable to some psychotherapy studies in the treatment of reducing stress, managing pain, and improving sleep (McCallie, Blum, & Hood, 2006). Other researchers have put forth how effective muscle relaxation is in the treatment of additional psychological disorders and physical conditions such as anxiety, mental distress, anger, chronic and acute pain (e.g., osteoarthritis and postsurgical), and cardiovascular rehabilitation (Rausch, Gramling, & Auerbach, 2006; Weber, 2004; Wilk & Turkoski, 2001). Freeman (2009) summed it up by stating that although there are clear design flaws in many of the available studies on progressive muscle relaxation many patients find that PMR is a beneficial intervention for the treatment of depression, anxiety, and other conditions that are heightened by mental anguish and stress.

According to Ghoncheh and Smith (2004) there are six major approaches to relaxation that health professionals can integrate into their work. They were identified as yoga, muscle relaxation (PMR), autogenic training, breathing exercises, meditation, and guided imagery. Thompson and Gauntlett-Gilbert (2008) explained that the general purpose of guided imagery is to induce a state of relaxation. Guided imagery is different from mindfulness in mindfulness has the goal of instilling a non-judgmental awareness in the present moment. Guided imagery actively seeks relaxation by focusing on the senses and the awareness of body movement (Meister et al., 2004). This particular CAM therapy involves the generation of mental images that utilizes imagination and visualization involving smell, touch, hearing, and taste. Detailed images are conjured up to bring forth a heightened state of awareness and relaxation and have been shown to be effective in the treatment and relief of various types of bodily pain (Park, 2013).

Guided Imagery

Guided imagery has been shown to be effective in the treatment of multiple mental and physical health conditions. Sloman (2002) conducted a study on advanced cancer patients and found that this complementary therapy helped the patients to have an overall higher quality of life and they experienced lower levels of anxiety and depression. Holmes and Metthews (2005) demonstrated in their research that guided imagery is effective in treating mood states and in regulating emotions. By employing the technique of imagery-rehearsal therapy Phelps and McHugh (2001) were able to show that those suffering from PTSD can be effectively treated for a range of symptoms and that the treatment also resulted in significant reductions in nightmares. These improvements were achieved by having the participants engage in pleasant imagery exercises and resulted in clear clinical improvements across a range of measures on PTSD and nightmares ($p < .0001$).

Guided imagery has also been studied across several other areas in the treatment of both mental and physical conditions and the data seem to indicate that its efficacy is well documented at this point. Some additional conditions that have been substantiated at this point include the treatment of addictions, grief, sleep disorders, phantom limb pain, asthma, immune conditions, and fibromyalgia (Forbes, Phelps, & McHugh, 2001; Menzies & Taylor, 2004; Wynd, 2005). Freeman (2009) summed up the importance of imagery by stating that this is the foundation of mind-body medicine and that it is the core element in many CAM therapies, including

biofeedback, relaxation therapy, hypnosis, and meditation.

Biofeedback

Another mind-body intervention that is gaining popularity and acceptance over time is

biofeedback. This CAM treatment involves the use of sophisticated technology and provides the

client with information on their physiologic states. Such states include galvanic skin

temperature, brain waves, muscle tension, and the goal is to empower the client in being able to

control these cognitive and physical states through deliberate mental control (Park, 2013). With

biofeedback training clients can also learn to alter their electroencephalographic (EEG) waves,

blood pressure, heart rate, and respiration (Moore, 2000). Biofeedback helps the individual to

gain control over their bodies (physiologic processes), behaviors, and cognitions.

The empirical support for the clinical efficacy of biofeedback at this point in time is

convincing. In fact, Park (2013) noted in her work that a number of researchers at this time

believe that biofeedback may actually be more of a conventional medicine due to the fact that its

methods and techniques are rooted in human physiology and is widely covered in mainstream

medical journals. Either way, at this point in time biofeedback is primarily regarded as a CAM

intervention. Several psychological and physical conditions appear to be effectively treated by

biofeedback, and they include anxiety disorders, depression, migraine headaches, chronic pain,

and cardiovascular disorders (Hammond, 2005; Hammond, 2007; Kaushik, Kaushik, Mahajan, &

Rajesh, 2005).

In her work on biofeedback research, Freeman (2009) put together a comprehensive list of

the various conditions that have been shown to be effectively treated with this mind-body intervention. In her analysis and critique of the literature Freeman stated that there is the least amount of empirical evidence to support claims that biofeedback is useful for treating the following conditions: autism, eating disorders, multiple sclerosis, and spinal cord injury. She concluded that were was modest empirical support that biofeedback is useful in treating alcoholism, substance abuse, insomnia, chronic pain, headache in children, irritable bowel syndrome, PTSD, traumatic brain injury (TBI), urinary incontinence in children, and other conditions. She concluded in her analysis of the relevant literature that biofeedback has the most empirical support and clinical efficacy in treating anxiety, hypertension, attention-deficit hyperactivity disorder (ADHD), headaches in adults, urinary incontinence in men, and temporomandibular disorders (Freeman, 2009). Based on the most recent research, biofeedback appears to be on the cusp of becoming more recognized as a traditional form of treatment and has a large amount of scientific evidence to support the claims concerning its clinical relevance and utility.

Hypnosis

Hypnosis is also considered an important mind-body intervention. Hypnosis involves intense concentration and focus with a certain level of suspension of awareness of the surrounding environment. It involves the absorption of the imagination and ideational experience, increased suggestibility, and cognitive dissociation (Park, 2013). Sharma and Kaur (2006) explained hypnosis as being a subjective state whereby the individual is at a heightened level of suggestibility. This altered state of consciousness is also accompanied by heightened perception and memory. The empirical support for hypnosis is strong in a couple of areas but it does not have the range of support as some of the other CAM interventions, such as biofeedback and yoga (Field, 2009).

In Freeman's (2009) analysis of contemporary research investigating the clinical efficacy of hypnosis it appears that the strongest evidence resides in the area of pain management. She went into detail on the benefits of hypnosis in the alleviation of pain across many areas, including cancer, spinal, surgical, and burn. She also highlighted research that demonstrated utility in the clinical treatments of obesity and post-traumatic stress disorder when used in combination with cognitive-behavioral therapy (CBT), pregnancy-induced nausea, smoking cessation, and phobias. Other researchers have shown the effects and use of hypnosis in clinical treatment in several areas including anxiety (Saadat et al., 2006), conversion disorder and PTSD (Jamieson, 2012), and irritable bowel syndrome (Whitehead, 2006). At this point in time it would appear that hypnosis is a useful intervention for treating the aforementioned psychological and physical conditions. Further research is needed in certain areas that hypnotherapists have claimed clinical utility and where the questionable quality of the research prevents definitive conclusions to be made (Park, 2013).

Yoga and Martial Arts

Yoga and martial arts have also been widely supported in the literature as effective CAM therapies across a wide range of conditions (e.g., Ghoncheh & Smith, 2004; Jamieson, 2012; Kaliappan, 1998; Telles & Naveen, 2004). These are the primary mind-body interventions detailed in this work and each will be closely connected to specific traditional therapies that are often employed in clinical practice today, namely cognitive-behavioral therapy and mindfulness-based interventions. The analysis and coverage on these specific practices are much more extensive than what has been covered on the other core mind-body therapies in use today (biofeedback, mediation, progressive muscle relaxation, etc.). As has been put forth in the coming sections of this work, it will be a central argument that yoga and martial arts are useful and effective complementary treatment modalities across a wide range of mental health conditions and behavioral problems.

Additional CAM Therapies

There are a number of other CAM therapies that are widely used in clinical practice today and that have strong support in the literature (Edenfield & Saeed, 2012; Park, 2013). Some of these complementary interventions include music therapy, chiropractics, homeopathy, aromatherapy, massage therapy, and acupuncture. There is ample support in the literature that these additional interventions also have utility in the treatment of psychological and behavioral conditions, such as stress reduction, anxiety, depression, nausea, substance abuse, and insomnia (e.g., Freeman, 2009; Karst, et al., 2007; White & Moody, 2006) but they have not been covered here as they are not typically regarded as mind-body or movement therapies, which is the primary focus of this dissertation. This work is focused on those interventions that involve both the mind and the body in the treatment of psychological and behavioral problems. These will be further examined in the upcoming sections of this work.

Chapter Three

Demonstration of the Empirical Evidence & Efficacy of CAM Therapies

The psychological literature is replete with efficacious explanations of the dominant therapies in use today and how they can be incorporated into treatment across the wide range of clinical disorders and behavioral problems most commonly seen in clinical practice. When it comes to combining the most dominant psychological therapies in use today with mind-body therapies there are two clear approaches that dominate the research and literature. As previously stated, cognitive–behavioral therapy and mindfulness-based interventions are the two therapeutic approaches that have been studied far more often with mind-body modalities than any other approaches in the field (Allison & Faith, 1996; Barnett, Shale, Elkins, & Fisher, 2014; Lau & McCain, 2005). These core therapies will be the central clinical approaches examined throughout this work and each will be analytically integrated with the respective complementary approaches of interest to this disquisition. The most relevant literature and research findings augmenting psychotherapy with CAM approaches will be cited.

Cognitive-Behavioral Therapy

Cognitive-behavioral therapy (CBT) has been studied more than any other traditional therapy when it comes to clinical integration with core CAM modalities. According to Plante (2011) cognitive-behavioral therapy is primarily based on the principles of learning and conditioning. Many of the techniques and principles of CBT are useful for treatments that are incorporating mind-body therapies into treatment (e.g., thought stopping, behavioral rehearsal, and counterconditioning). Spiegler and Guevremont (2010) noted that the core tenets of CBT involve the modification of cognitions directly through cognitive interventions and indirectly through behavioral interventions. These interventions often include restructuring therapy where the client learns to change distorted and erroneous thinking through learning new coping skills, through self- talk, and through mastery training.

As an example of how traditional CBT interventions can be used in combination with mind-body interventions, Spiegler and Guevrmont (2010) explain how thought stopping works and why it is useful in helping to modify the client's unhealthy thinking. Thought stopping is a way to terminate irrational or defeating thoughts by interrupting the problematic patterns in one's thinking. Negative thoughts are replaced with more positive and productive thoughts in a conscious and deliberate manner (Plante, 2011). Thought stopping goes hand in hand with the principles of certain mindfulness and meditative practices. This technique, as well as other CBT techniques, can be used in conjunction with certain mind-body approaches to help the client realize a more efficacious outcome.

Cognitive restructuring is another traditional approach that can be used combination with certain CAM modalities. Lau and McMain (2005) put forth that empirical evidence supports the position that cognitive restructuring serves to reduce recurrence or relapse of depression. This is accomplished by facilitating the decentering and awareness of negative thought patterns. Decentering basically involves any technique that is aimed at altering centered thinking toward a more open and inclusive state of mind (American Psychological Association, 2007). Cognitive techniques mindfulness techniques both address issues dealing with feelings, physical sensations, thoughts, so it makes sense that CBT would be a good fit for certain CAM therapies. With this stated, there are also differences that clinicians should be aware of (Thompson & Gauntlett-Gilbert, 2008). Mindfulness approaches tend to emphasize an awareness in thinking similar to cognitive interventions but there are differences in regard to how one should make sense of, and deal with, transient mental events. According to Beck (1995) the cognitive approach holds that effective change resides in realistic evaluation and thinking. This priority placed on changing the content of thinking and private experience is necessary in the modification and improvement of behavior. Mindfulness approaches do not view these unwanted private experiences as necessarily harmful or problematic (Baer, 2003). Although there are some differences between the two approaches they can be used in combination to further assist the client in his or her healing.

Cognitive-behavioral training has also been shown to be effective in combination with muscle relaxation therapy in the treatment. Progressive muscle relaxation has been shown to be one of the more efficacious CAM modalities for treating certain mental health and physical issues (Granath, Ingvarsson, von Thiele, & Lundberg, 2006). Jones and Johnston (2000) concluded that although cognitive-behavioral skills training is more effective in the treatment of psychological issues than progressive muscle relaxation training, when used in combination this has been shown to be more effective across a ranges of outcomes measures than using any single technique (e.g., the treatment of depression and stress management). Van der Klink, Blonk, Schene, and Van Dijk (2001) supported this position in their work covering cognitive-behavioral interventions with meditation, biofeedback, and muscle relaxation where they concluded that cognitive-behavioral techniques alone were the most effective in treating certain psychological problems (e.g., depression and stress), but in combination with these specific CAM modalities the outcomes were significantly improved. The empirical evidence put forth in the literature seems to clearly support the integrationist approach being endorsed in this work.

Chen (2011) also supported the integration on cognitive-behavioral therapy with certain mind-body approaches. In his work he demonstrated that CBT can be effectively combined with relaxation techniques, meditation, guided imagery martial arts and yoga to treat both mental and physical ailments (e.g., recovery from stroke, delirium, confusion, disorientation, etc.). He argued that integrating mind-body modalities with CBT is growing in popularity due to the minimal side effects (if any), convenience, low cost, and usefulness. This growing popularity can be expected to continue and will only become a more central part of Western medicine and treatment (Barnett, et al., 2014; Chen, 2011; Field, 2009).

Several studies have been published that have demonstrated the usefulness in combining cognitive-behavioral therapy with hypnosis in the treatment of eating disorders, obesity, and weight control. Freeman (2009) covered a range of studies looking at how integrating hypnosis with CBT therapy can substantially enhance treatment outcomes for the treatment of obesity. She looked at several meta-analyses on this matter and determined that clients that undergo this method of treatment show continued improvements over time in weight management and improved eating behaviors. Kirsch (1996) performed several meta-analyses on the effects of integrating hypnosis with CBT techniques and concluded that the benefits increased considerably over time and was particularly useful for long-term management of weight loss and eating behaviors (effect size = 0.98, $r = 0.74$). As an example, he found that the average weight loss for CBT alone was around 6 pounds, but when combined with hypnosis it increased an average of almost 15 pounds. Similar efficacious findings have also been found in studies looking at combining hypnosis with traditional therapeutic interventions (e.g., Jamieson, 2012; Sadaat et al., 2006).

Cognitive-behavioral therapy has also been used in conjunction with biofeedback to treat chronic pain. Morley, Eccleston, & Williams (1999) conducted a meta-analytic review across 25 randomly controlled trials that cognitive-behavioral treatments and biofeedback produced significantly improved results across all measures that what was observed in control groups. The outcome measures in this comprehensive analysis included cognitive coping, mood, affect, cognitive appraisal, and pain experience. Chambless and Ollendick (2002) also found that treatment groups that received behavioral therapy with biofeedback had significantly improved results over control group treatments. In other words, it was concluded by the authors that chronic pain was better treated by this approach than by standards treatments alone. It would appear from the current empirical research that combining cognitive and behavioral interventions with biofeedback is an efficacious approach in the treatment of bodily pain, coping management, and mood regulation (Morley et al., 1999).

Cognitive-behavioral therapy has also been combined with certain mind-body interventions to treat anxiety disorders (e.g., PTSD). Schoenberger (2000) concluded that the data at the time of his analysis showed promising results for the treatment of anxiety disorders, pain management, and obesity through CBT and hypnosis but he was tentative due to the limitation of randomly controlled trials utilizing this particular therapeutic combination. With this said, Kirsch, Montgomery, and Sapirstein (1996) found that CBT combined with hypnosis resulted in a substantial effect size in the treatment of anxiety, insomnia, obesity, and pain across multiple measures. In fact, the researchers concluded that CBT in combination with hypnosis resulted in a 70% improvement in patients compared to when CBT was used without hypnosis. Even though the empirical support for these treatment combinations has often been complicated in the past by less than ideal standardized procedures and techniques, the clinical evidence for its use and efficaciousness is mounting over time as more robust studies surface each year (Field, 2009).

Mindfulness and Mindfulness-Based Interventions

The use of mindfulness-based interventions (MBI's) in psychological treatment has become more popular in recent years (Crane, 2009; Thompson & Gauntlett-Gilbert, 2008). The use of MBI's is increasingly being used to reduce problem behaviors, to improve social skills, to achieve overall wellness, to relieve anxiety, to treat learning disabilities, to treat ADHD, and so on (Haydicky et al., 2012; Krisanaprakornkit et al., 2006; Mace, 2008). Edenfield and Saeed (2012) noted in their work that studies focusing on MBI treatments are increasingly being published in psychological literature for both mental and physical illness. The authors pointed out in their work that mindfulness-based interventions have been increasingly demonstrated to effectively treat a range of psychological disorders, such as depression, anxiety, and a range of stress-related conditions.

Mindfulness-based interventions are rooted in Buddhist and yoga philosophies. Mindfulness refers to purposefully being in the present moment in a nonjudgmental way and accepting way (Kabat-Zinn, 2003). This ideally involves being fully aware and in touch with what is taking place in the present moment. Martin (2011) described mindfulness practice as the devotion of intentional attention to what is revealed to us through directed focus toward experience of one's internal and external world. The goal is to deliberately attend to the experience itself that involves being engaged, observing, noticing, and being aware in a nonjudgmental way. This is related to Yalom's (2005) understanding of the *mindfulness of being* whereby the individual is aware of oneself in an authentic state that embraces potential and accepts limitations. It is through this authentic mindful state of being where one accepts *that* things are and by being aware of one's self-creation they can obtain the power to positively change their lives. Martin (2011) also promoted the belief that being mindful can help to improve one's life in a number of ways. He also argued that being mindful in therapeutic practice can be very beneficial to the therapist while serving the client.

Mindfulness-based interventions have been shown to be useful in the treatment of both adult and children across a wide range of conditions. Adkins, Singh, Winton, McKeegan, and Singh (2010) studied a mindfulness-based procedure known as *Meditation on the Soles of the Feet* to demonstrate that unhealthy thinking and maladaptive behavior can be effectively treated by shifting attention form the problem emotion (e.g., fear, anger, frustration, etc.), or any other trigger, to a neutral object (such as the soles of the feet). They demonstrated in their research that this mindfulness-based procedure is useful in the treatment of aggression and anger management, as well as treating those with intellectual disabilities and mental illness (e.g., depression, mental retardation, obsessive-compulsive disorder, etc.). Other studies have also demonstrated the clinical efficacy of mindfulness training and interventions in the treatment of maladaptive behaviors and mental illness (Bogels e al., 2008; Haydicky et al., 2012).

Goyal et al. (2013) conducted a meta-analytic review of 47 randomized clinical trials with 3,515 participants that used mindfulness meditation in the treatment of several mental health conditions. They concluded that mindfulness programs moderately improved anxiety (effect size = 0.38), depression (effect size = 0.30), and pain (effect size = 0.33). They also concluded that these practices were not significant (low or insufficient evidence of effect) in improving sleep, eating habits, weight, or substance use. They also did not find strong evidence to support the notion that these mindfulness practices are any better at treating these disorders/conditions any better than active and traditional treatments (e.g., behavioral therapies, exercise, drug therapy).

With this stated, the researchers also expressed support for these mindfulness-meditative

programs in the clinical treatment of multiple dimensions of psychological stress (Goyal et al.,

2013). Other research teams have found that mindfulness practices are beneficial in the

treatment and management of stress and mood states. For example, Caldwell, Harrison, Adams,

Quin, & Greeson (2010) found that mindfulness interventions improve self-regulatory self-

efficacy which has the effect of improving negative arousal, mood, stress, and relaxation. The

researchers concluded that increased mindfulness improves sleep quality and overall quality of

life.

In a meta-analysis on 21 adult mindfulness studies, Baer (2003) found that chronic pain,

depression, and anxiety were effectively treated with mindfulness interventions (Cohen's $d =$

0.74, SD $= 0.39$). Although not all of the studies were randomly controlled trails there was still

sufficient evidence to support the position that mindfulness practices can be used in

psychological treatment to improve the lives of clients. Baer (2003) concluded by stating that

MBI's can likely bring those suffering from mild to moderate psychological distress back into

(or close to) the normal range of functioning. Edenfield and Saeed (2012) put forth in their work

that MBI's are effective in the treatment of stress, anxiety, and depression. They argued that the

mechanisms of change that MBI's produce will typically result in an overall improvement in

general well-being and health. On this point they put forth:

Improved understanding of the role of MBI's in affecting positive changes on correlates

and precipitants of psychological disorders (e.g., rumination, negative automatic

thoughts, emotional reactivity) is important in improving the translation of research to

practice, specifically regarding the treatment and prevention of impairing conditions such as recurrent and chronic psychological disorders (e.g., major depressive disorder, generalized anxiety disorder). (Edenfield & Saeed, 2012, p.136)

The current literature is replete with research and clinical trials that have demonstrated the use and effectiveness of treating mental and behavioral conditions with mindfulness-based interventions. This work has placed its attention on the two dominant approaches in use today – mindfulness-based cognitive therapy (MBCT) and mindfulness-based stress reduction (MBSR). Both of these approaches are rooted in Buddhist and yoga traditions and they have both become widely used treatments in modern clinical practice (Barnett et al., 2014; Freeman, 2009; Lau & McMain, 2005; Spiegler & Guevremont, 2010). Mindfulness-based therapy was originally developed to treat depression (Spiegler & Guevremont, 2010) but has since also shown therapeutic efficacy in others areas as well (Coelho, Canter, & Ernst, 2013). Mindfulness-based cognitive therapy and mindfulness-based stress reduction have both demonstrated clinical utility across a range of studies that involved exercise and bodily movements, such as tai chi, yoga, karate, and aikido (e.g., Granath, Ingvarsson, von Thiele, & Lundenberg, 2006; Lengacher, 2012). The use and effectiveness of each of the MBI approaches in combination with mind-body therapies has been elaborated on extensively in the forthcoming sections of this work.

Mindfulness-Based Cognitive Therapy

Mindfulness-based cognitive therapy (MBCT) was developed with the intention of preventing relapse or recurrence of major depression. Even though MBCT is a relatively new "third wave" cognitive-behavioral therapy it has demonstrated promising clinical utility across a range of psychological issues such as suicidal behaviors, anxiety, disorders, insomnia, binge eating, and bipolar disorder (Coelho, Peter, Canter, & Ernst, 2013; Spiegler & Guevremont, 2010). This approach to treatment is grounded in Buddhist and yoga traditions and places a heavy emphasis on mindfulness meditation. Mindfulness meditation is important to the practice of MBCT because it helps the client to acknowledge, in a nonjudgmental manner, distracting thoughts and feelings. This philosophy assists the individual in detachment from those unwanted or harmful thoughts and feelings and to gain a deeper understanding and awareness that allows them to be more flexible and adaptive in their coping responses (Edenfield & Saeed, 2012). One of the key distinctions between traditional cognitive-behavioral therapy and mindfulness-based cognitive therapy is the emphasis placed on acceptance (as opposed to change). This awareness and attitude of acceptance of personal experience is what allows the individual to be experientially open to their present reality (Lau & McMain, 2005).

Mindfulness-based cognitive therapy is naturally a good fit for complementary therapies. One of the reasons for this is because of how the sessions are structured and what is emphasized in the work. A typical MBCT session includes a small group that is taught mindfulness meditation techniques that focus on inner thoughts, feelings, and physical sensations (recall that meditation is a core mind-body therapy) and last for about 2 to 3 hours. The group meetings are held once per week for 8 weeks (Field, 2009). Barnett et al. (2014) posited that contemporary research indicates that it is the integration of mindfulness meditation with traditional cognitive-behavioral approaches that serves to make MBCT so effective in the treatment of depression and in the prevention of depressive relapse. According to Lau and McMain (2005) it is this connection to mindfulness that allows the client to accept their experience and to embrace the Zen philosophy that true freedom can be achieved through nonattachment to experience and developing an attitude of acceptance.

As described in the earlier section covering mindfulness meditation as a CAM therapy, MBCT also integrates the philosophy that clients must learn to change the focus of their attention and how they make sense of their experience and bodily reactions. One of these methods is accomplished through "decentering," whereby the individual is taught to disconnect themselves from their thoughts and to realize that these thoughts do not represent what they are (Spiegler & Guevremont, 2010). This emphasis on mindfulness training teaches clients how to disengage from frames of mind that are characterized by harmful and ruminative thinking and to adopt a new mindset of acceptance and one of "being" (Coelho, Canter, & Ernst, 2013). Lau and McMain (2005) proclaimed that this focus on mindfulness meditation in MBCT allows the cultivation of an attitude of acceptance that has a major impact on the quality of the therapeutic relationship, the client's acceptance of self and others, and the clinician's acceptance of the client. It is through these acceptance experiences and mindful practices that one will come to realize positive behavioral and attitudinal changes (Lau & McMain, 2005).

The goal of MBCT is to change the awareness of and reaction to thoughts. The goal is not to change thought content as is typically done in CBT therapy. The belief is that cultivating a decentered connection to negative thinking will allow one to prevent the escalation on negative thinking during times of potential relapse (Lau & McMain, 2005). This is the foundation of MBCT and the empirical evidence to date on its ability to significantly reduce depressive relapse or recurrence is promising (Field, 2009). The fundamental belief is that major life stressors are the strongest predictors of first onset of depression than of recurrent episodes. Dysphoric mood and negative thinking styles are more closely associated with having a history of depressive episodes and relapses (Lewinsohn, Allen, Seeley, & Gotlib, 1999).

Core goals of MBCT are to reduce environmental stressors and the ways that one responds to them by increasing awareness of negative thinking styles during times of potential relapse (Teasdale, Segal, & Williams, 1995). It is the associations between depressed mood and negative thinking styles that lead to an increased likelihood of reactivation of harmful thinking that can lead to further depressive states. Thus, the goal is to reduce the risk of relapse or recurrence by changing the patterns in cognitions that become disruptive in dysphoric mood states and to normalize these cognitive patterns so they are manageable and do not escalate (Lau, Segal, & Williams, 2004). Although the data looks promising for MBCT in the treatment of depression and reducing relapse, Field (2009) pointed out that it is still not exactly clear how mindfulness mediation facilitates traditional cognitive-behavioral therapy and that it has primarily been the developers of MBCT that have published the merits of this approach. The author also noted that MBCT still needs to be clinically compared with other forms of treatment in regard to its effectiveness in reducing depressive episodes (Field, 2009).

There have been numerous studies conducted that have demonstrated the efficacy of MBCT in treating patients across a range of populations and conditions. For example, Lengacher (2012) reviewed the effectiveness of mindfulness-based cognitive therapy in the treatment of depression with cancer survivors. Several hundred cancer survivors were treated with MBCT and the author makes a strong claim that the approach is clinically effective in helping cancer survivors in their deconstruction of what took place in their lives, in their suffering, and in their recovery and movement toward a more positive state of mind. Yook et al. (2008) demonstrated in their research that MBCT was effective in treating insomnia across a range of patients suffering from anxiety disorders. Williams et al. (2008) put forth in their work that MBCT is even showing promise in the treatment of bipolar disorder.

One of the strongest studies conducted on MBCT in the treatment of depression came from the work of Ma and Teasdale (2004). In this randomized controlled study they demonstrated that MBCT was especially effective in preventing relapse and/or recurrence in clients that have had three or more depressive episodes. The results of their research showed that MBCT reduced relapse in clients from 78% to 36%. Similar beneficial findings have been reported in a number of additional studies (e.g., Manicavasagar, Parker, & Perich, 2011; Veehof, Oskam, Schreurs, & Bohlmeijer, 2011). A primary focus of this dissertation is grounded in demonstrating the therapeutic efficacy of combining MBCT with particular mind-body therapies in the treatment of certain mental health conditions and behavioral problems. The goal of this section of the work was to put forth the merits of MBCT in therapeutic practice. The relevant literature strongly indicates that mindfulness-based cognitive therapy is a useful addition to the clinician's therapeutic skill set (e.g., Coelho et al, 2013; Lau et al., 2004; Ma & Teasdale, 2004).

Mindfulness-Based Stress Reduction

Mindfulness-based stress reduction (MBSR) is also a core mindfulness-based intervention that has been intensively studied for its efficaciousness in treating certain mental health conditions. MBSR has also been well researched in combination with certain CAM therapies and the evidence to date is promising. According to Lothes, Hakan, and Kassab (2013) MBSR is an effective technique that teaches people to be more in touch with the present moment and aware of themselves and their experiences. Park (2012) explained mindfulness-based stress reduction as structured meditation that is rooted in Buddhist principles. It was originally developed as a group-based program for those suffering from chronic pain and involves specific techniques such as body scans, seated meditation, and yoga. Edenfield and Saeed (2012) also point out that this clinical approach to treatment is also heavily grounded in yoga philosophies and techniques. At the core of this approach is learning to focus one's attention toward what is taking place within oneself at the present moment. This internal focus is done in a nonjudgmental way and allows the individual greater awareness and insight into their negative or distracting thoughts (Kabat-Zinn, 2003).

In a similar manner the mindfulness-based cognitive therapy, MBSR also works toward cultivating a focus on ongoing subjective experience that is combined with an attitude of openness and awareness to experience Edenfield & Saeed, 2012). According to Freeman (2009) MBSR is effective in the treatment of stress reduction with nonclinical populations and in stress management. She went on to state that although more research is needed with clinical populations, MBSR has great potential as an intervention in both clinical and medical settings for the treatment of pain, depression, and psychosocial adaptation (e.g., functional disability, psychiatric symptoms, and emotional distress; Freeman, 2009). Other studies have supported the use of MBSR across a range of psychological and medical conditions, such as in the treatment of anger/aggression, depression, anxiety, confusion, and cancer (Adkins, et al., 2010; Bishop, 2002; Speca, Carlson, Goodey, & Angen, 2000). Although MBSR was originally conceived as a treatment for stress and related conditions, the empirical evidence is mounting that is also beneficial for a wide range of mental health conditions such as ruminative thinking, psychological distress, mood disturbance, poor attention, improved sleep/insomnia, cognitive disorganization, emotional irritability, anxiety, etc. (Brown, Ryan, & Creswell, 2007; Garland, 2007; Lee et al., 2007; Shigaki, Glass, & Schopp, 2006).

Mindfulness-based stress reduction has also been demonstrating clinical utility in the treatment of medical conditions coupled with psychological distress. Research into MSBR has shown great promise in helping those suffering from cancer and recovery from cancer (Freeman, 2009). Carlson, Speca, Patel, and Goodey (2003) studied the relationship between MBSR treatment for those suffering from breast cancer and prostate cancer and stress symptoms, quality of life, and mood states. The team studied 49 patients with breast cancer and 10 patients with prostate cancer across an 8 week MBSR program that included the core techniques of this approach (e.g., meditation, yoga, and relaxation). They found that after the treatment there was a clear shift in immune profiles (e.g. through measurement of the production of certain types of cells and chemicals) and that the patient's had significant improvements in symptoms of stress, sleep quality, and overall quality of life. Their general conclusion was that MBSR is an effective mental health and medical treatment for those fighting cancer (Carlson et al, 2003).

In a subsequent study investigating the use of MBSR in the treatment of 59 patients with breast cancer and 10 patients with prostate cancer, Carlson, Speca, Patel, and Goodey (2004) found similar findings through measurements of plasma DHEAS, salivary cortisol, and salivary melatonin, and they once again concluded that MBSR is an effective approach for improving patient's stress symptoms, mood states, and overall quality of life. As in their earlier study (Carlson et al., 2003), the research team again included the traditional MBSR techniques of yoga, relaxation, and meditation. Here we can see that combining MBSR with certain mind-body therapies (e.g., yoga and meditation) results in efficacious and beneficial results for those being treated. Other studies have also demonstrated the benefits of using stress reduction techniques and meditation in the treatment of medical conditions couples with mental health problems, such as treating cardiac conditions , hypertension, and immune conditions (e.g., Ditto, Eclache, & Goldman, 2006; Lane, Seskevich, & Peiper, 2007; Otto et al., 2006).

Some of the more interesting findings concerning MBSR and how it improves lives come from neuropsychological and medical research. There has been strong evidence that MBSR can improve memory, learning, emotional regulation, self-referral processing (Goldin & Gross, 2010). In their research the team found that MBSR literally promotes structural and functional changes in the brain (e.g., amygdala activity and gray matter concentration) and serves to improve certain mental health conditions and cognitive functioning, such as anxiety, depression, attention, and focus. Additional studies in these areas have shown that MBSR also results in improvements and changes in the autonomic nervous system (e.g., blood pressure, heart rate, etc.), on alpha and theta brainwave activity, and on the thalamus, prefrontal cortex, cerebral cortex, and anterior cingulated cortex (e.g., Freeman, 2009; Orme-Johnson, Schneider, Son, Nidich, & Cho., 2006).

Based on the current literature is would appear that MBSR is an effective treatment approach across a range of psychological and medical conditions (Field, 2009; Goldin & Gross, 2010). The core ingredients in mindfulness-based stress reduction come from yoga and Buddhist philosophies and this is what seems to set this approach apart from more traditional psychological therapies (Edenfield & Saeed, 2012; Park, 2012). The evidence to date is very convincing that MBSR is a useful therapeutic modality and that in combination with certain mind-body therapies it efficaciousness is even further enhanced (Carlson et al, 2004; Lykins, E. L., & Baer, 2009; Park, 2012). The therapeutic integration with specific CAM modalities will be further explored in this work. As of this writing the empirical evidence for using the specific mindfulness-based interventions covered in this work (i.e., MBCT and MBSR) for the treatment of psychological and physical conditions seems convincing. This integrative potential with specific mind-body therapies will be further explored and reinforced across this comprehensive review.

Chapter Four

Yoga as a Complementary Mind-Body Therapy

Yoga Therapy

One of the central purposes of this endeavor is to research the merits of yoga as a complementary form of therapy to be used in conjunction with more traditional psychological therapies. With this stated, there is growing base of literature that proposes that yoga in and of itself is valid form of therapy. Although the goal of this work is to research the benefits of combining certain mind-body therapies (i.e., yoga and martial arts) with more mainstream psychological therapies it is still important to understand the foundations and merits of these adjunct approaches as stand-alone therapies. As an example of this growing movement, the International Association of Yoga Therapists, which views yoga as a healing art and science and publishes the *International Journal of Yoga Therapy*, has as its central mission the objective of establishing yoga as a recognized and respected therapy (IAYT, 2014).

Yoga therapy (YT) is the modus of empowering the individual to make progressive steps toward improved health and well-being through the application of the philosophy and methods of yoga. In yoga therapy the focus is on adapting the techniques of yoga to address and alleviate the specific concerns of the individual (Kaley-Isley, Peterson, Fischer, & Peterson, 2010). Yoga therapy is a flowing movement exercise that, in addition to having a positive impact on the physical body, enhances mental concentration, reduces cortisol levels, increases vagal activity, and improves cognitive performance (Field, 2009).

Yoga therapy is a form of mind-body medicine that has been practice to promote relaxation, physical fitness, and mental health and wellness for around 5,000 years (Barnett et al.,

2014) and has been shown to be effective in alleviating a wide range of psychological and physical health issues, such as depression, anxiety, ADHD, stress, hypertension, heart disease, headaches, diabetes, asthma, etc. (Michalsen et al., 2005; Wolsko, Eisenberg, Davis, & Phillips, 2004). YT is typified by a series of poses that are called *asanas* in Sankrit and each involves slow concentrated movements and deep abdominal breathing (Field, 2009). YT is viewed by practitioners as a structured process that is designed to purify the body and mind from toxins that have been acquired through improper lifestyle choices and harmful thinking patterns (Bhatia et al., 2012). This goal is accomplished through the techniques of meditation, deep breathing exercises, and postures designed to improve the body's energy system. Therapeutic yoga has as its primary goal the intention of alleviating suffering, facilitating mastery, and improving the quality of life for practitioners (Kaley-Isley et al., 2010).

According to Bhatia et al. (2012), YT has been demonstrated to have beneficial effects on perceptual sensitivity, memory functions, and cognitive remediation for those with impaired cognition (e.g., psychiatric illnesses). Ghoncheh & Smith (2004) argued that yoga is therapeutic in the sense that it alone can improve cognitive and somatic symptoms in clients. In support of this position, Culos-Reed, Carlson, Daroux, and Hately-Aldous (2006) found in their study that YT had both physical and psychological benefits for breast cancer survivors. In their investigation the team randomly assigned eligible participants to either the intervention group (n = 20) or the control group (n = 18). After conducting pre- and posttests the findings demonstrated that YT was effective in treatment and fostered improvement in patients in the areas of emotional functioning, global quality of life, tension, mood disturbance, cognitive disorganization, and depression. The findings also showed that YT was helpful in treating certain physical problems such as diarrhea and gastrointestinal symptoms (Culos-Reed et al., 2006).

Similar findings were found in Levine and Balk's (2012) meta-analytic study of 71 articles looking at the effects of yoga therapy on breast cancer patients. Although these investigators found that many of the studies analyzed would require greater methodological rigor to be able to better describe the mechanisms responsible for the effects, there is suggestive evidence that YT improves patients overall quality of life (i.e., emotional health, social functioning, physical well-being, and functional adaptation; Levine & Balk, 2012). Park (2012) proposed in her work that YT is useful for the treatment of a variety of mental and physical health conditions, such as diabetes, stress management, asthma, and even adverse effects of aging. Researchers have also put forth that yoga therapy has been demonstrated to be effective in the treatment of chronic pain and resultant mental health problems (Park, 2012; Williams et al., 2005). In the Williams et al. (2005) study looking at chronic low back pain it was shown in a randomized control trial that after 16 weeks of YT significant improvements were found in functional disability and pain intensity. The participants in the study also reduced their usage of pain medications by 88% and showed a clear improvement in psychological health and general attitude.

Yoga therapy has also been shown to have a biological effect such as the reduction of cortisol levels in the brain. West et al. (2004) found that YT produced a significant main effect for salivary cortisol which they concluded was directly associated with reductions in stress and negative affect. They proposed that the neuroendocrine response to YT resulted in a significant interaction influencing positive affect in the participants. Field (2012) addressed something similar in her work where she proposed that YT lowers anxiety levels by lowering the electrodermal activity in the brain. This, in turn, results in lower cortisol levels and an improved affective state. Kaley-Isley et al. (2010) supported this view that yoga therapy is effective is lowering salivary cortisol levels and decreasing anxiety in children and adolescents. They proposed that this physiological response to YT is associated with both stress reduction and positive affect which in turn serves to alleviate anxiety.

Some research has looked deep into how YT can alter brain chemistry and be used to treat mental illnesses that have largely been reserved for psychotropic treatment. Although there have been questions raised in regard to the methodological rigor of some of these research studies (e.g., Cramer, Lauche, Klose, Langhorst, & Dobos, 2013) there does appear to be promise and utility (e.g., Bhatia et al., 2012). In the Bhatia et al. (2012) research they found through multiple comparisons that YT as an adjunctive therapy had a significant effect on attention and cognitive function in patients with schizophrenia (n = 65). They also found that the improvements were greater among the men than among the women (although the reason for this is not clear). The findings concluded that there were nominal effect sizes resulting from YT intervention and this also held for patients that have bipolar I disorder (n = 40) and depressive disorder (n = 37; Bhatia et al., 2012). Varambally et al. (2012) demonstrated through a randomized controlled study that YT improves negative symptoms in those with schizophrenia compared to controls. These findings held that YT increases functioning in those with schizophrenia several times beyond those that received other forms of alternative treatment.

In another randomized control trial looking at the effects of YT on those with schizophrenia it was found that this intervention resulted in significant improvements in psychological well-being, paranoia, energy levels, depression, and thought disturbance (Visceglia & Lewis, 2013). The researchers concluded that YT reduced both the positive and negative symptoms of schizophrenia in the patients. With this said, Cramer et al. (2013) put forth that there are still several matters that need to be clarified in regard to YT being used as a

treatment for schizophrenia. Although they acknowledge that there are signs of promise in

regard to improving cognitive function and psychological well-being, they argue that research to

date often conflates findings with antipsychotic medications, lacks methodological structure in

treatment protocols, and seems to have a pattern of bias in the reporting of findings (Cramer et

al., 2013). With this stated, researchers do acknowledge that a number of studies using YT

interventions have followed rigorous methods and have upheld the scientific standards of

randomization, using control groups, and have followed tight statistical controls (e.g., Bhatia et

al., 2011; Cramer et al., 2013; Williams et al., 2005).

It has been put forth here that yoga therapy as a stand-alone intervention can be useful in

the treatment of a wide range of psychological conditions and physical health problems. This

has been accomplished by pulling in numerous studies that have looked at the clinical

effectiveness of YT across a diverse range of populations and settings. For the purposes of this

work the overarching goal is to demonstrate how yoga, as a complementary mind-body therapy,

can be used in conjunction with more traditional psychological therapies. It is believed that to be

able to better understand the benefits of these clinical pairings it is essential to have a

fundamental understanding of how and why yoga therapy works as an independent intervention.

As stated by the International Association of Yoga Therapists, yoga therapy is both an art and a

science that deserves to be recognized and respected as an effective form of therapeutic treatment

(IAYT, 2014).

Chapter Five

The Clinical Efficacy for Specific Psychological Problems

Relevant Psychological and Behavioral Issues

Complementary and alternative therapies have been studied across a wide range of psychological and behavioral conditions. The empirical evidence to date appears to show that various CAM modalities are effective in treating some of the most common psychological disorders presented by those seeking professional assistance (Barnett & Shale, 2012; Basler, 2011; NCCAM, 2010). It is not the goal of this work to put forth that complementary mind-body therapies are optimal, or even desirable, for all psychological or behavioral conditions. This comprehensive effort has as its goal the bringing forth and illumination of those symptoms and conditions that have been demonstrated to be clinically ameliorated through the use of the specific CAM modalities included in this analytic review. Although CAM therapies have been shown to be clinically useful for treating a wide range of physical conditions and diseases (e.g., musculoskeletal, neurological, autoimmune) the focus of this piece will revolve primarily around certain psychological disorders and behavioral issues.

Depression. Mood disorders are among the most studied psychological conditions when it comes to CAM therapies (Astin, 1998). The evidence to date suggests that depression is one of the leading psychological disorders that are the most effectively treated through the use of certain mind-body therapies (Freeman, 2009). In her work covering certain mind-body therapies and depression, Freeman argued that this mental health condition can be effectively treated using meditation, relaxation therapy, exercise therapy, and massage therapy. In regard to meditation she put forth that meditation alone is an effective mood modulator and can be used as a treatment for both emotional and mental dysfunction (Freeman, 2009).

Many other researchers have also put forth their evidence that depression can be effectively treated using certain CAM therapies. Examples of studies that have shown the therapeutic benefits of mind-body therapies in the treatment of depression include the modalities of yoga (Michalsen et al. (2012), acupuncture (Wang & Kain, 2001), meditation (Veehof et al., 2011), music therapy (Hsu & Lai, 2004), and biofeedback (Karavidas et al., 2007). It would appear from the literature that exercise therapy has been demonstrated to be one of the most effective and productive methods for treating depression (e.g., Arent, Landers, & Etnier, 2000; Harris, Cronkite, & Moos, 2006; Singh, Clements, & Fiatarone-Singh, 2001). The range of studies investigating the usefulness in treating depression with exercise techniques is extensive and this appears to be one of the most highly recommended CAM therapies for this particular mental health problem (Field, 2009).

Depressive disorders were put forth as a major public health problem by Martinsen (2008). In his work he argued that changes in lifestyle are one of the best ways to prevent and treat depression. He proposed that exercise, as a supplement to traditional therapy, is the best way to truly treat depression over the long-term. His findings put forth that physically active people are at a significantly reduced risk of developing depression and that various exercise therapies can significantly alleviate mild to moderate depression (Martinsen, 2008). Along with her recommendation of employing exercise in the treatment of depression Field (2009) also put forth that this particular psychological condition can also be effectively treated using a range of other CAM modalities, including massage therapy, tai chi, yoga, music therapy, mediation, and biofeedback.

In their meta-analysis of the clinical benefits of employing CAM therapies into treatment of mental health and physical conditions, Barnett et al. (2014) question the use of certain mind-body therapies in the treatment of depression. Although they acknowledge that these modalities have become very popular, they also point out that many of these approaches need more controlled studies to truly demonstrate treatment efficacy. They argued that one of the main problems is the relative paucity of randomized controlled trials (RTCs). They highlighted one of the main problems as residing in the inability of CAM researchers to be able to create sham and control conditions. They used the modalities of acupuncture and massage to demonstrate why some conditions are so difficult with certain (not all) mind-body therapies. Without being able to adopt the gold standard in research (RTCs) CAM researchers are in a double bind. They wrote that, "Without the ability to adequately control or blind to condition, CAM researchers are faced with a substantially greater burden of proof for their findings than would a study of a pharmacological treatment regimen" (Barnett et al., 2014, p. 35). With this said, the authors did ultimately put forth in their work that certain CAM therapies do show promise in the treatment of depression.

Field (2009) and Freeman (2009) showed a very different attitude toward mind-body therapies in the treatment of depression and extensively articulated their beliefs that the empirical support for clinical efficacy at this point is quite strong. Kaley-Isley et al. (2010) also summarized their research by stating that certain CAM therapies (e.g., yoga) have demonstrated efficacy in the treatment of depression and that CAM therapies in general are legitimate interventions in the treatment of a wide range of mental and physical health conditions. With this information presented, the evidence seems to be mounting that the core mind-body therapies are useful and effective in the treatment of clinical depression (Harris et al., 2006; Hsu & Lai, 2004).

Anxiety. A number of CAM therapies have been shown to be effective in treating anxiety disorders. According to Kessler, Soukup, and Davis (2001) anxiety is one of the most commonly studied mental disorders when it comes to treatment through CAM modalities. Although not all of the core mind-body therapies have been demonstrated to be efficacious in treating anxiety disorders a handful of them are showing great promise and have been widely supported in the literature. Freeman (2009) put forth that exercise, meditation, and massage therapy have the most support in the research at this point in time. She did highlight hypnosis as a showing promise in the improvement of phobias and reduction of anxiety.

A number of other authors have also put forth exercise as one of the best CAM modalities for treating anxiety (e.g., Barbour, Edenfield, & Blumenthal, 2007). Chesney et al. (2006) argued that engaging in at least 30 minutes of moderate-intensity exercise (e.g., martial arts) 5 days per week will significantly improve psychological well-being and will reduce feelings of anxiety. Wang (2012) also argued that martial arts, as a form of exercise, will reduce anxiety (as well as depression and stress). In a comprehensive review of the effects of exercise in reducing anxiety Martinsen (2008) found that a regular exercise regimen, in combination with traditional therapies, is one of the most effective ways to prevent and treat anxiety. Freeman (2009) reviewed several studies that looked at the effects of single bouts of exercise in producing anxiolytic effects as well as the impact of regular high-intensity workouts. In comparing control groups across studies to those groups that engaged in the exercise routines she concluded that exercise does have anxiety-reducing effects.

Meditation is one of the core CAM modalities that appear to have among the strongest support when it comes to treating anxiety. Krisanaprakornit et al. (2006) conducted a review of clinically randomized trials in the treatment of anxiety disorders through meditation practices. They determined that meditation as a therapy for anxiety has the general effect of reducing arousal and of ameliorating symptoms across different anxiety conditions. Goyal et al. (2013) conducted a meta-analytic review of 47 clinical trials and determined that mindfulness meditation does often result in moderate levels of improved anxiety and lower levels of psychological distress. Edenfield and Saeed (2012) analyzed the usefulness of mindfulness meditation in the treatment of anxiety. They found that effects sizes were robust across multiple independent investigations. In a meta-analytic review they concluded that mindfulness practices produced significantly greater results that placebo conditions (Hedges' $g = 0.45$, 95% CI: 0.35-0.46). The researchers stated that although the underlying mechanisms of change need to be further studied the evidence to date clearly shows that meditation has beneficial effects in the treatment of anxiety (Edenfield & Saeed, 2012). Freeman (2009) generally supported this position by stating that practicing meditation on a regular basis reduces anxiety for those with clinically elevated levels of anxiety. She also pointed out that meditation, as a primary intervention, may initially be less effective in treating those with long-term and severe anxiety disorders because these clients may have difficulty focusing. Specific clinical limitations regarding mindfulness meditation in the treatment of anxiety were also pointed out by Toneatto and Ngyuen (2007).

Yoga appears to be among the most effective when it comes to the treatment of anxiety disorders. In fact, in virtually of all of the research collected for this work yoga was almost universally supported when it comes to treating this particular mental health problem (e.g.,

D'Angelo, 2013; Edenfield & Saeed, 2012; Field, 2009; Kuan-Yin et al., 2011; Park, 2012; Salmon, Lush, Jablonski, Sephton, 2009). Kuan-Yin et al. (2011) concluded in their meta-analytic review of relevant randomized control trials that yoga produced significant effects sizes in the treatment of anxiety. Kaliappan (1998) used yoga in his work with 200 adolescent boys and found that this mind-body therapy, in combination with behavior modification techniques, resulted in reduced social anxiety. D'Angelo (2013) found that this CAM therapy was clinically useful in the treatment of anxiety for certain populations (e.g., female, those that felt greater responsibility for their health). Khalsa, Greiner-Ferris, Hoffman, and Khalsa (2014) employed yoga in a clinical intervention with 32 participants suffering from generalize anxiety disorder and concluded that there were significant clinical improvements across the sample by the end of the six weeks of treatment.

Other mind-body therapies have also been extensively researched. Massage therapy was supported as an effective CAM intervention for the treatment of anxiety by Barnett et al. (2014). This research team did recommend certain mind-body therapies for the treatment of anxiety (e.g., mediation, yoga, massage therapy) but they did not support the use of others (e.g., acupuncture, hypnosis, music therapy). Field (2009), the director and founder of The Touch Research Institute (which studies the effects of message therapy on clinical populations) and one of the leading researchers and proponents of this particular CAM modality posited that massage therapy works on anxiety by lowering cortisol levels, reducing tension, inducing a relaxed state, and by improving overall mood. Freeman (2009) also brought forth the evidence that massage therapy reduces salivary cortisol levels, induces relaxation, and reduces anxiety. She highlighted that massage therapy alters EEG patterns and brain wave activity in a desired manner and thus decreases anxiety and anxious behavior. In a meta-analytic review of the benefits of treating anxiety with massage therapy, Moyer, Rounds, and Hannum (2004) concluded that the evidence is convincing at this point that this particular CAM therapy is effective in the treatment of anxiety.

In her summary concerning which of the CAM therapies have been demonstrated to be clinically useful in the treatment of anxiety she put forth the additional modalities of martial arts, acupuncture, progressive muscle relaxation, and imagery. Although other researchers agree with her position on the various CAM approaches in the treatment of anxiety (Hammond, 2005; Lee et al., 2007), others do not (Barnett et al., 2014). Park (2012) supported the use progressive muscle relaxation and biofeedback in the treatment of anxiety. She noted that both of these approaches have been shown to be effective in helping patients to control tension and to obtain mental calmness. Freeman (2009) stated that the research is mixed when it comes to progressive muscle relaxation, but that overall this approach does not seem to reach the desired results. Barnett at al. (2014) determined that although the scientific evidence at this point is not consistent regarding the use of biofeedback in the treatment of anxiety, there have been several respected randomized control trials that have produced telling results. They stated that the evidence to date is strongest for employing biofeedback in the treatment of performance anxiety for sports and music.

Although Barnett et al. (2014) stated that evidence is weak in regard to acupuncture, aromatherapy, imagery, and hypnosis, and that they do not recommend using these CAM practices for the treatment of anxiety, Field (2009) disagreed and stated that these are useful complementary treatments and have been shown to be effective for treating this particular condition. The evidence at this point in time seems to suggest that certain CAM modalities are clinically effective in the treatment of anxiety (e.g., exercise, meditation, yoga, massage) some are debated among researchers (e.g., biofeedback, progressive muscle relaxation, martial arts,

acupuncture), and others have not been supported to any extent in the scientific literature that

would merit discussion here (e.g., music therapy, Pilates, hypnosis, Reiki, aromatherapy).

Stress. The reduction of stress and stress management are among the top reasons why

people seek out CAM interventions for psychological conditions (Park, 2012). There are several

CAM therapies that stand out in the literature in regard to the treatment of stress and stress-

related conditions. One of the mind-body therapies that appear to have strong empirical support

is mindfulness meditation. Mindfulness interventions such as meditation were among the first

mind-body therapies to be by mainstream healthcare providers in the United States and they have

been shown to be clinically effective in the treatment and reduction of stress and stress-related

symptoms (Caldwell et al., 2010; Park, 2012; Thompson & Gauntlett-Gilbert, 2008). Edenfield

and Saeed (2012) stated that the research data have consistently shown that mindfulness

meditation improves emotional well-being and quality of life by, in part, significantly reducing

stress. Goyal et al. (2014) support this claim in their research and found that meditation

generally helps to improve mental health and quality of life and that it can alleviate

psychological stress in those that commit to the practice.

Mindfulness-based stress reduction (MBSR), as discussed in an early section of this work,

was designed with a primary intent of reducing stress and stress-related symptoms. The

literature is replete at this point on the merits of MBSR in the treatment of psychological stress

and this approach to treatment is now widely accepted for this purpose (e.g., Carlson, Speca,

Patel, & Goodey, 2004; Edenfield & Saeed, 2012; Lin, Hu, Chang, Lin, & Tsauo, 2011). In their

meta-analytic research on the therapeutic effects of MBSR, Chiesa and Serretti (2009) found that this approach to treatment clearly results in reductions in stress and overall improvements in quality of life.

Yoga is another mind-body intervention that appears to have among the strongest support in the prevailing literature in regard to stress management and stress reduction (e.g., D'Angelo, 2013; Field, 2012; Ghoncheh & Smith, 2004; West, Otte, Geher, Johnson, & Mohr, et al., 2004). Edenfield and Saeed (2012) described the effects of yoga on stress by pointing out how yoga teaches slow and deep breathing strategies that alleviate psychological distress by balancing both sympathetic and parasympathetic responses. The authors went on to say that yoga works by effectively governing stress-related networks in the body and by inducing a state of relaxation. Field (2012) explained how and why yoga is effective in reducing stress in the practitioner. She claimed that yoga lowers both cortisol levels and levels of electrodermal activity, and thus, effectively calms the individual and reduces stress. Ghoncheh and Smith (2004) argued that yoga is the most popular method of relaxation among the general public and that its ability to calm the individual has been widely examined in both research and practice. In connection to yoga and stress management in a clinical setting, cognitive-behavioral therapy has also been shown to be a useful therapeutic approach in the treatment of stress management when used in combination with this particular CAM intervention (Ingvarsson, Thiele, & Lundeberg, 2006).

Martial arts have also been shown to be an effective mind-body intervention for the treatment of stress and stress management (Wang, 2012). Certain practices in the martial arts have demonstrated to be beneficial in calming the mind, in teaching the practitioner how to control their breathing, and in lowering blood pressure (e.g., aikido, tai chi, chi gong, and jujitsu). The outcome of this calming of the mind and body results in reduced stress and higher cognitive functioning (Gemmell & Leathem, 2006; Wang, 2011). The effects of martial arts training on stress levels have been measured across various instruments (e.g., Cohen's Perceived

Stress Scale) and the data is convincing that these mind-body therapies have positive benefits

when it comes to stress reduction and stress management (Taylor-Piliae et al., 2006; Park, 2012).

Mind-body interventions are continuing to demonstrate valid clinical use for stress

management and stress reduction as more controlled studies come forth in the literature. Field

(2009) analyzed the effectiveness of mind-body therapies in the treatment of stress and she

concluded that the evidence at this point is quite convincing that massage therapy, exercise,

music therapy, aromatherapy, progressive muscle relaxation, imagery, meditation, and

biofeedback are all valid and useful techniques for this particular need. In regard to

aromatherapy and music therapy, Field put forth that most of the research to date has revolved

around the treatment of stress and that the evidence is sound and that they both have useful

benefits. Freeman (2009) supported these claims and stated that both aromatherapy and music

therapy have psychologically calming effects that alters body functioning and lowers stress

levels. Ghoncheh and Smith (2004) backed the recommendations made by Field in regard to

progressive muscle relaxation and proposed that this form of therapy has been shown to calm

patients and reduce stress across multiple measures (e.g., Smith Relaxation States Inventory).

Park (2012) also supported the use of PMR and biofeedback in the treatment of stress. Barnett et

al. (2014) disagreed with the above proclamations and declared that the evidence in the literature

is weak regarding the effects of music therapy, aromatherapy, imagery, PMR, and biofeedback.

The only two interventions that they supported were massage therapy (strong recommendation),

and MBSR (mindfulness meditation). The research team did not analyze the current literature in

regard to exercise as a mind-body therapy for the treatment of stress (Barnett et al., 2014).

Overall, and based on the meta-analytic reviews that have been forth, it would appear that meditation, yoga, massage therapy, different forms of exercise (e.g., aerobic, swimming, running, weight lifting, etc.), and certain forms of martial arts (e.g.,tai chi, qi gong, aikido, etc.) have the most support at this point in time when it comes to treating stress (e.g., Field, 2009; Moraska, Pollini, Boulanger, Brooks, & Teitlebaum, 2010; Taylor-Piliae et al., 2006; Wang, 2012). Progressive muscle relaxation and biofeedback has moderate support in the literature at this time and the evidence seems to be mounting on its effectiveness in treating stress and in stress management (e.g., Freeman, 2009; Park, 2012; Pawlow & Jones, 2005). The other CAM therapies do not have the level of empirical support when it comes to treating stress in psychological practice and in controlled trials. Mindfulness-based stress reduction and cognitive-behavioral therapies have also been shown to be clinically effective in the treatment of stress and stress management (e.g., Chiesa & Serretti, 2009; Hu et al., 2011; Ingvarsson, 2006) and this will be explored further in combination with certain CAM therapies in a later chapter in this work.

Attention deficit/hyperactivity disorder. Certain complementary mind-body therapies have been shown to have beneficial effects on attention-deficit/hyperactivity disorder (ADHD). As a disorder that typically has an early onset and that is characterized by marked inattention and poorly overactive modulated behavior ADHD can be effectively treated by meditation practices (Krisanaprakornkit, Ngamjarus, Witoonchart, & Piyavhatkul, 2010). The authors reported in their research that mindfulness meditation has demonstrated in several randomized controlled trials (n = 49) to be as effective in treating this condition as standard therapy and drug therapy. Although they propose that this CAM intervention has benefits in treating this particular mental health disorder further clinical trials are needed. Although Freeman (2009) did not put forth evidence for meditation in the treatment of ADHD she did examine how the relaxation effects of massage therapy have been shown to reduce cortisol levels, improve concentration, and decrease anxiety in children and adolescents with this developmental condition. Field (2009) also supported the use of massage therapy in treating ADHD, particularly in pediatric populations.

Traditional treatment for ADHD usually involves behavioral and pharmacological interventions but the use of CAM therapies is on the rise across the United States. This increasing use of CAM modalities in the treatment of this developmental condition is particularly true for children and teenagers ages 3 through 17 (Muir, 2012). Biofeedback appears to be one of the core mind-body therapies that have the most support in the literature when it comes to treating ADHD. Numerous studies utilizing widely accepted measures and testing (e.g., theta-beta waves, cortical potentials, Attention Deficit Disorders Evaluation Scale, Test of Variables of Attention, etc.) have shown that biofeedback is an effective technique when working with ADHD (e.g., Arns, de Riddler, Strehl, Breteler, & Coenen, 2009; Holtman & Stadler, 2006; Monastra, 2005). After reviewing the relevant clinical trials that have been done with biofeedback in the treatment of ADHD, Barnett, et al. (2014) gave a strong recommendation for utilizing this approach. They stated in their analysis that the empirical evidence to date strongly indicates that this CAM intervention is a beneficial technique when treating this condition. According to Freeman (2009) EEG biofeedback has the greatest application when it comes to the treatment of ADHD.

Yoga and martial arts have also been widely supported in the literature in regard to working with children and adolescents with ADHD. Haydicky, Wiener, Badali, Milligram, and Ducharme (2012) found in their research with 60 adolescents with ADHD that martial arts training over a 20-week period resulted in significant improvements anxiety levels, attention, and behaviors. Palermo et al. (2006) also found in their research that martial arts training decreases noncompliance in children with ADHD and clinically reduces the range of symptoms associated with it. Field (2012) put forth that both yoga and martial arts are useful CAM practices in the treatment of this ADHD. She found in her research that martial arts (e.g., tai chi) helps to treat ADHD by lowering stress hormones, reducing anxiety, heightening alertness, and by reducing hyperactivity.

In regard to the merits of yoga, Field (2012) described how yoga is one of the most effective CAM therapies when it comes to treating ADHD. She put forth that the current research on this mind-body practice demonstrates that yoga markedly decreases attention deficits and enhances attentiveness due to its long periods of mental concentration. Yoga also works with ADHD by lowering anxiety and cortisol levels (Field, 2012). Kaley-Isley et al. (2010) conducted a meta-analysis on the effects of yoga in the treatment of ADHD and found that the evidence shows that this CAM practice is a beneficial adjunctive therapy to standard care and that it clearly serves to improve symptoms and behaviors. They analyzed both randomized and nonrandomized research covering a wide range of research with children and adolescent populations and found that yoga consistently demonstrated to be effective across different validated tests and instruments (e.g., Conners Rating Scale) and consistent findings were demonstrated between trial groups and comparison groups (Kaley-Isley et al., 2010).

Sleep/Insomnia. There has also been a sizable amount of research looking at the benefits of certain CAM therapies in the treatment of sleep disorders/problems and insomnia. Fouladbakhsh and Stommel (2010) conducted a secondary analysis of the 2002 National Health Interview Survey with a sample of 2,262 adults aged 18 years and older that have been diagnosed with cancer. They found in their secondary analysis that CAM practices are widely used among women that have been diagnosed with cancer and that the integration of self-care CAM interventions is effective in symptom management (e.g., fatigue, pain, depression, and insomnia). The researchers reported that yoga, certain styles of martial arts, guided imagery, and meditation are all effective in alleviating sleep problems and insomnia. Barnett et al. (2014) stated in the meta-analytic review that insomnia is one of the commonly cited reasons for clients seeking out complementary and alternative therapies. Overall it would appear that certain CAM modalities are useful in the treatment of sleep problems and insomnia. Field (2009) gave strong support for the use of massage therapy, yoga, aromatherapy, music therapy, and tai chi in the treatment of insomnia and sleep disturbances/problems.

The literature appears to be mixed in regard to the benefits of mindfulness-based practices. Caldwell et al. (2010) examined mindfulness practices with 166 college students over a 15-week period and found that tiredness, negative arousal, relaxation, and stress were all alleviated. The end result was that the college students generally experienced improved sleep quality and overall quality of life. Goyal et al. (2013) found the opposite in their research that included 47 clinical trials. They determined that mindfulness meditation practices showed little evidence for improving quality of sleep. Field (2009) also did not recommend meditation for sleep improvement or for treating insomnia due to the low level of quality research available at this time.

One of the most widely supported CAM interventions for the treatment of sleep disturbances and insomnia is yoga (e.g., Field, 2009; Kozasa et al., 2010; Levine & Balk, 2012). Khalsa (2004) put forth after conducting a primary study that yoga improved sleep on virtually every measure (e.g., sleep time, sleep quality, sleep efficiency, sleep onset latency, number of awakenings, etc.) and found that this mind-body therapy was clearly effective in the treatment of chronic insomnia. Levine and Balk (2012) reviewed 71 articles from seven databases and ultimately determined from their analysis that yoga was among the most beneficial mind-body therapies for improving quality of life (QOL) and improving quality of sleep for patients undergoing treatment for breast cancer. Kozasa et al. (2010) conducted a PubMed search and found 12 randomized controlled trials where the main objective was to treat insomnia with mind-body interventions. They concluded that yoga and similar relaxation techniques were clinically useful in the treatment of insomnia. They also determined that cognitive-behavioral therapy was the only intervention that produced better results than medication across the trials. In a unique manner, they classified and put forth CBT as mind-body therapy and positioned it as the most effective mind-body intervention for treating insomnia (Kozasa et al., 2010).

Martial arts as a mind-body therapy have also been supported in the current literature in regard to treating sleep disturbances and insomnia (e.g., chi gong & tai chi; Field, 2009; Fouladbakhsh & Stommel, 2010; Wall, 2005). Kozasa et al. (2010) conducted a meta-analysis of 12 randomized controlled trials published in PubMed and were resolute in their conclusion that tai chi can ameliorate sleep quality, improve total sleep time, and improve sleep efficiency. Fouladbakhsh and Stommel (2010) were also decided in their position that tai chi and chi gong (a.k.a. qi gong) were effective CAM practices in alleviating pain and measurably reducing insomnia in patients with cancer. In her meta-analytic work looking at the use of complementary therapies in clinical practice, Field (2012) was firm in her stance that tai chi improves sleep and alleviates sleep disturbances. She elucidated that the prevailing research clearly demonstrates that this style of martial arts improves pulmonary functions, increases vagal activity, and decreases stress hormones, thus leading to better quality of sleep and fewer sleep disturbances.

Similar findings regarding the benefits of tai chi and sleep quality were also reported in research by Wall (2005) and Wang et al. (2004).

Healing from trauma. Traumatic experiences can lead to a range of mental and behavioral health problems. Certain CAM therapies and traditional therapies have been shown to be effective in the treatment of mental and physical pain and injury due to traumatic experiences. Yoga is one of the mind-body therapies that have been demonstrated to be useful when it comes to treating pain and injury from trauma (Kaley-Isley et al. 2010). The authors explained how yoga has been used to treat anxiety and depression associated with pain and injury through specific mind-body movements where participants move gently in and out of poses with awareness of effects, directed attention, visualized breathing, the releasing of pain, and the observing of any changing sensations. Barnett et al. (2014) reviewed several studies that demonstrated that yoga serves to alleviate bodily pain and mental distress.

Veterans suffering from post-traumatic stress disorder (PTSD) have demonstrated significant clinical improvements after participating in VA treatment programs utilizing yoga and meditation (Libby, Reddy, Pilver, & Desai, 2012). The researchers claimed that the evidence is substantial at this point that yoga therapy is an effective therapeutic discipline and that it should be included in VA patient-centered care models with veterans that are suffering from PTSD (and other mental health issues). Dhikav et al. (2010) conducted a 12-week yoga program for women (ages 22-55) that were suffering from various sexual dysfunctions and sexual disorders rooted in different causes such as sexual assault, poor body image, early trauma, etc. They used the Female Sexual Function Index (FSFI) scale to measure any improvements in the participants after the yoga training. They found improvements were found in all six domains of the FSFI (e.g., pain, orgasm, lubrication, satisfaction, desire, and arousal; $p < 0.0001$).

Although Field (2009) put forth that yoga is useful for treating certain issues dealing with pain and injury and associated mental health conditions (such as anxiety and depression) she placed her emphasis on the merits on massage therapy. Out of the core CAM therapies identified in the literature massage therapy was the only one that she endorsed for the treatment of sexual abuse. Freeman (2009) supported the use of massage therapy for a range of traumatic experiences including trauma and pain from burns, from surgery, from cancer treatment. Freeman also put forth that cognitive-behavioral therapy and hypnosis have been shown to be useful in healing and recovery from traumatic experiences. Kubsch, Neveau, & Vandertie (2000) found in their research on patients that have suffered from severe injuries or trauma that those treated with massage therapy showed markedly reduced symptoms and improvement related to pain, heart rate, blood pressure, mood, and anxiety.

Martial arts as a mind-body therapy have also gained a respectable amount of support in the literature. Guthrie (1995) argued that martial arts therapy is one of the best ways to heal the mind and body following traumatic experiences. She and her research team worked with hundreds of women that have been the victims of sexual abuse, incest, rape, and domestic violence and found that martial arts therapy was the best way to get the women in touch with both their minds and bodies and typically results in significant overall healing, as well as marked improvements in empowerment, body image, self-esteem, and in promoting an overall positive sense of self. Jamieson (2012) reviewed the benefits of including CBT and tai chi in the treatment of those suffering from PTSD and concluded that the findings are promising regarding its clinical effectiveness. Tai chi has also been reported to positively assist with coping and overall functioning in those dealing with post-surgery trauma (Mustian, Katule, & Zhao, 2006). The marital art style of qigong has been reported to be a useful mind-body therapy in the recovery, health, and coping of those suffering from the effects of chemotherapy. This has been most widely studied with women undergoing treatment for breast cancer (Yeh, Lee, Chen, & Chao, 2006).

Other CAM therapies that are showing promise in clinical treatment of pain, trauma, PTSD, and healing include biofeedback, physical exercise, and music therapy (Barnett et al., 2014; Hernandez-Ruiz, 2005). In her work with abused women that were living in shelters, Hernandez-Ruiz (2005) found that music therapy was an effective method in reducing stress and anxiety and that it also helped the women to regain mental calmness and improve their quality of sleep. Barnett et al. (2014) reviewed the current literature and concluded that biofeedback has shown clinical utility in the treatment of veterans suffering from PTSD. The research team also proposed that music therapy might be uniquely suited to treat PTSD and that the evidence to date is promising. In a randomized treatment intervention utilizing music therapy, Carr et al. (2012) demonstrated that patients that did not respond to traditional CBT treatment did show a significant reduction of PTSD symptoms and had marked improvements over those assigned to the control group.

Aggression/externalizing behaviors. Certain mind-body therapies have also been shown to be very successful in the treatment of aggression and externalizing behaviors. The merits of these CAM interventions appear to be especially true regarding treatment with children and adolescents (e.g., Bogels, Hoogstad, van Dun, de Schutter, & Restifo, 2008; Palermo et al., 2006). Across the range of CAM modalities that have been comprehensively examined in regard to the treatment of aggressive and externalizing behaviors, martial arts have been the most researched and empirically supported. Haydicky et al. (2012) integrated mindfulness martial arts with cognitive-behavioral therapy in the treatment of adolescents with externalizing behavior, oppositional defiant and conduct problems. They demonstrated after a 20-week mindfulness training program that externalizing behavior and social skills were clinically improved based on parent ratings scores and symptom analyses.

One of the primary reasons that martial arts has been studied in relation to aggressive and externalizing behaviors is because there is strong evidence that it can help to reduce or eliminate harmful and antisocial behaviors (Palermo et al., 2006). This group of researchers argued that childhood and adolescent disruptive behaviors can often lead to serious deviance and noncompliance with rules and regulations later in life and that martial arts training may be one of the best ways to intervene and treat these behavioral problems. In their study the researchers randomly assigned children to an experimental group that went through a 10-month program in *Wa Do Ryu* karate. The results showed clinically significant improvements in scale scores compared to controls. The team concluded that this systematic karate intervention demonstrated clear efficacy in the treatment of these children with externalizing conditions and disruptive behaviors by substantially reducing (improving) their problem behaviors (Palermo et al., 2006).

Other researchers have also put forth the merits of martial arts in the treatment of aggressive and externalizing behaviors. In making their case for the use of martial arts in treating individuals that act out aggressively (even violently), Twemlow, Sacco, and Fonagy (2008) argued that these types of individuals are often resistant to abstract words and ideas (e.g., as with psychodynamic psychotherapy) and therapists often cannot seem to "reach" them. With this, they put forth that martial arts therapy embodies the process by using physical movement and by addressing the kinesthetic core of early attachments and aids them in re-tooling their experiences. There basic position was that aggressive, even violent, nonmentalizing patients usually do not respond well to talk therapy approaches and this is why physically oriented therapies (such as martial arts and yoga) are valuable complements. A range of researchers have also supported the use of martial arts for treating aggressive and violent behaviors, such as tai chi, karate, qigong, jujitsu, judo, and aikido (e.g., Kerr, 2002; Lantz, 2002; McKenna, 2001). The empirical evidence to date appears to show that martial arts training as a mind-body therapy is one of the most effective ways to treat these particular behavioral problems in children, adolescent, and adults (Haydicky et al, 2012; Palermo et al., 2006).

Other CAM therapies have also received support in the extant literature. Field (2009) gave her support to massage therapy and progressive muscle relaxation in the treatment of aggressive behaviors. Kaliappan (1998) treated 35 delinquent boys with behavioral therapy, systematic desensitization techniques, and yoga over a 4 month period and found that negative attitudes, social anxiety, social deviance, and disruptive behaviors all significantly reduced. The author also put forth another sample finding with 200 adolescents deviant males where yoga in combination with behavior modification techniques clinically reduced deviant and disruptive behaviors and also improved overall academic performance (Kaliappan, 1998).

Adkins et al. (2010) combined behavioral interventions with mindfulness meditation and found in their research that this therapeutic combination resulted in statistically significant improvements in verbal aggression, physical aggression, and a range of disruptive behaviors. Bogels et al. (2008) treated adolescents with a range of externalizing disorders (e.g., conduct disorder, oppositional defiant disorder) and found that after 8 weeks of mindfulness training that impulsivity, externalizing, withdrawal, self-control, and attention problems were all clinically improved. They proposed that those adolescents with behavioral control deficits and externalizing disorders can be effectively treated through clinically controlled mindfulness-based behavioral training (Bogels et al., 2008). The evidence for the therapeutic merits of these specific mind-body therapies (e.g., martial arts, yoga, and mindfulness meditation) in the treatment of aggression, externalizing behaviors, and conduct disorders is well supported in the current literature and continues to be reinforced through empirical research (e.g., Adkins, 2010; Kerr 2002, Palermo et al., 2006).

Eating disorders/weight issues/obesity. The literature also seems to lend support for the use of certain adjunctive therapies in the treatment of eating disorders and weight issues. Several studies have combined hypnosis with and cognitive-behavioral therapy the general findings were that this is an effective way in enhance treatment outcomes (e.g., Allison & Faith, 1996; Kirsch, Montgomery, & Sapirstein, 1995). In their meta-analytic review of using CBT with hypnosis, Kirsch et al. (1995) concluded that the evidence suggests that using this combination of treatments results in greater long-term outcomes and weight loss. In a subsequent meta-analysis, Kirsch (1996) demonstrated that both the effect sizes and correlations (0.98 and 0.74, respectively) showed that using hypnosis as a supplemental therapy with cognitive-behavioral interventions has clinical utility and tends to result in greater benefits than any one approach alone.

Yoga has also been shown to be beneficial in the treatment and management of eating disorders and weight issues. Kaley-Isley (2010) reviewed several studies that evaluated the merits of using yoga as a complementary therapy for children and adolescents with eating disorders and weight problems. They concluded that the outcomes measures (e.g., from the Eating Disorder Examination/EDE inventory) all supported the position that those youths that included yoga into the treatment had significantly reduced body mass index levels, reduced food preoccupation, and reduced impulsive eating behaviors. Carei, Fyfe-Johnson, Breuner, and Brown (2010) reached the same conclusions when they studied the value of incorporating yoga into the treatment of eating disorders with adolescents and young adults. They found in their randomized control trial that scores on the EDE showed clear improvement in the participants and that this CAM modality has clinical utility in the treatment of eating disorders. Field (2012) also supported the use of yoga in treating children and adolescents with yoga and found in her research that this mind-body practice does help with decreasing body fat and improving chronic diseases associated with being overweight or obese. She found that yoga decreased eating disorder symptoms by reducing the drive for thinness, improving body image and body satisfaction, and by reducing symptoms of depression and anxiety.

In her research on the most effective ways to decrease body fat, to treat eating disorders, and to improve the health of children and adolescents, Field (2012) put forth that aerobic exercise was one of the single most useful mind-body interventions. Her evidence suggests that intense to moderately intense exercise programs have shown the overall best outcomes in regard reduced body mass index, reducing blood pressure, decreasing post-exercise heart rate, improving oxygen consumption, and getting obese youths back within normal to healthy ranges. Freeman (2009) also supported the inclusion of exercise as an adjunctive therapy in the treatment of body weight issues and improved behavioral control, as well as for the treatment and facilitation of cognitive functioning, improving mood states, controlling stress levels, etc. Freeman's work found support for using this CAM approach for treating individuals from all age cohorts – including children, adolescents, middle-aged adults, and aging populations.

In her comprehensive work looking at the most commonly practice CAM therapies, Field (2009) recommended that aromatherapy, imagery, and exercise have all shown empirical efficacy when it comes to the treatment of weight issues and obesity. In this same work her strongest support for the adjunctive treatment of eating disorders was given to massage therapy.

She explained that massage therapy serves to decrease body dissatisfaction, lowers stress and anxiety, increases dopamine levels, and lowers cortisol levels, which all assisted in the treatment of anorexia nervosa. Field also concluded from her research that massage therapy successfully treats bulimia by lowering depression scores, lowering cortisol levels, increasing dopamine levels. Her findings indicated that participants showed overall improvements in impulsive behaviors and on eating disorder scales (Field, 2009).

Addiction. Several CAM therapies have demonstrated promise when it comes to the treatment and management of addiction. Hypnosis is one of the CAM modalities that have been researched the most when it comes to smoking cessation. Elkins, Marcus, Bates, Rajab, and Cook (2006) demonstrated in a randomized study that hypnosis was as effective as pharmacological and more traditional behavioral methods in helping clients to stop smoking. After 26 weeks of treatment they had a smoking cessation rate of 40%, which they attributed to the hypnotic intervention. Green and Lynn (2000) conducted a meta-analysis on the efficacy of hypnosis a treatment for smoking cessation and concluded that the evidence is clear that is does have clear effects on reducing rates of smoking compared to control treatments and wait-list groups.

In another systematic investigation on the benefits of using hypnosis in the treatment of smoking addiction, Tahiri, Mottillo, Joespeh, Pilote, and Eisenberg (2012) found in their meta-analytic review of randomized controlled trials and concluded that hypnosis was an effective way to quit smoking and that the success rates of hypnosis were just as high as they are with pharmacological treatments. Freeman (2009) is a bit more reluctant in her assessment of hypnosis as an effective treatment and stated that reports are mostly anecdotal and that hypnosis alone does not provide a cure and is best used in combination with other treatment modalities (e.g., behavioral). Barnes et al. (2010) conducted a meta-analysis of randomized control trials on the use of hypnotherapy in treating smoking addiction and found that there are no significant differences between hypnotherapy and traditional psychological therapies.

Acupuncture has also shown promise as a CAM modality in the treatment of addictions. White and Moody (2006) demonstrated in a meta-analytic review that acupuncture was an effective treatment for smoking addiction, while other researchers have put forth that acupuncture has clinical utility in the treatment of substance abuse and addiction involving drugs such as opiates and cocaine (e.g., Gurevich, Duckworth, Imhof, & Katz, 1996; Lipton, Brewington, & Smith, 1994). Other researchers have conducted systematic reviews of the literature regarding the use of acupuncture in the treatment of drug addiction and have concluded that no significant differences are noticeable when comparing treatment group to control group outcomes (e.g., Jordan, 2006).

Other CAM modalities have been researched for their efficacy in the treatment of addictive behaviors (e.g., drug use and smoking). Field (2009) gave her support to guided imagery, aromatherapy, and massage therapy when it comes to smoking cessation and addiction to nicotine. Although she lent her support for these interventions she was still clear that the number of studies in these areas is still scant and that more controlled research is needed. Freeman (2009) did not provide support for the aforementioned CAM therapies put forth by Field but she did position meditation as a mind-body intervention that is showing promise in the treatment of addictive behaviors. Freeman (2009) noted that meditation has been shown to be useful in drug abuse outcomes by creating a more optimal homeostasis, reducing stress, increasing serotonin availability, increasing tryptophan, and by reducing cortisol levels. She concluded by stating that the evidence seems to indicate that meditation is an effective intervention for decreasing drug abuse and preventing abuse of chemical substances.

Other mind-body therapies have also been supported in the literature. Guthrie (1995) found in her work with abused women that martial arts assist in the treatment and recovery from drug abuse by providing women with the tools they need to take control of both their minds and their bodies. Martial arts are also believed to be effective for treating behavioral issues (e.g., drug use and addiction) by empowering individuals through a corrective emotional experience that strengthens and improved self-concept, self-control, and discipline (Twemlow, 2008). Biofeedback is another complementary approach that has also shown promise in the treatment of substance abuse and addiction. For example, in a randomized trial investigating the effectiveness of biofeedback in the treatment of cocaine addiction, Burkett, Cummins, Dickson, and Skolnick (2005) demonstrated that utilizing this CAM modality resulted in a clinically significant reduction in relapse in the treatment group compared to relapse rates in the control group (51% relapse compared to 70% relapse, respectively). The research to date appears to demonstrate by and large that these specific mind-body therapies have a place in the treatment of harmful addictive behaviors. This clinical utility appears to only be more pronounced when combined with traditional psychological interventions (e.g., Freeman, 2009; Tahiri et al. 2012).

Chapter Six

Yoga and Martial Arts as Core Mind-Body Therapies

Relevant Mind-Body Therapies

The empirical evidence to date seems to lend an abundance of support for the use of complementary therapies in the treatment of a wide range of psychological and behavioral issues. This work does not favor any one approach over another and acknowledges equal value to each of the core mind-body therapies that have been well established in the literature. Several examples of CAM therapies that are relevant to this work and that have been widely examined for their effectiveness in psychological treatment include, but is not limited to, biofeedback (e.g., Barnett et al., 2014; Karavidas et al., 2007), meditation (e.g., Manicavasagar et al., 2011; Veehof et al., 2011), hypnosis (e.g., Barnes et al., 2010; Tahiri et al., 2012), massage therapy (e.g., Freeman, 2009; Moraska et al., 2010), yoga (e.g., Field, 2012; Kaley-Isley et al., 2010), progressive muscle relaxation (e.g., Ghoncheh & Smith, 2004; Weber, 2004), acupuncture (e.g., Jordan, 2006; White & Moody, 2006), and martial arts (e.g., Field, 2012; Palermo et al., 2006). Although all of the CAM therapies covered in this work are valuable and relevant, specific attention will be given to two mind-body therapies of particular interest: yoga and martial arts.

Yoga. Yoga is a form of mind-body medicine that has been practiced for around 5,000 years and is most commonly used to improve and maintain psychological well-being, mood, and physical fitness (Brotto, Mehak, and Kit (2009). According to Park (2012), the intention of yoga is to instill complete health and balance to the practitioner. This goal is primarily accomplished through stretching exercises and poses/postures that develop strength and flexibility, meditation, breathing exercises, and sustained concentration (Khalsa et al., 2014).

Yoga follows the practice and belief of *sutras*, or the eight foundations (limbs). In brief the eight yoga sutras are, yama (moral behavior), niyama ((healthy habits and tolerance), asana (different physical postures), pranayama (breath control/breathing exercises), pratyahara (withdrawal of senses and detachment), dharana focused attention/deep concentration), dhyana (contemplation/meditation), and Samadhi (achieving higher consciousness, supreme bliss, and enlightenment; Barnett et al., 2014; Park, 2012).

According to Brotto et al. (2009) one of the main reasons that yoga has increased in popularity over the years is because of its positive effects on mental health and functioning. They proposed that it is through yoga's ability to balance the *chakras* that energy is channeled in a healthy and constructive manner and the restless mind is calmed. This calming of the mind and redirecting of energy and mental focus serves to alleviate psychological suffering and improves the practitioner's behaviors, attitudes, thoughts, and perceptions (Brotto et al., 2009; Kaley-Isley, 2010). The benefits of employing yoga in the treatment both mental and physical conditions has been widely explored and documented (e.g., D'Angelo, 2013; Fouladbakhsh & Stommel, 2010; Libby et al., 2012; Levine & Balk, 2012). Wolsko et al. (2004) put forth a wide range of psychological and medical conditions that they found are treated or alleviated through the use of yoga, such as anxiety, depression, hypertension, back pain, stress, ADHD, cancer treatment symptoms, and heart disease.

The most common schools of yoga practiced in the United States are Hatha (slower paced and focused on breathing and stretching), Iyengar (focus on physical postures and body alignment, relaxation, and breathing), and Bikram (use of 26 postures in high-temperature rooms). Regardless of the particular form employed, all traditions of yoga are designed to improve the practitioner's mental and physical state (Barnett et al., 2014). Field (2009) spelled out how the typical yoga session is structured. The practitioner wears nonrestrictive clothing and has bare feet. The exercises consist of a series of poses (*asanas*) that are performed slowly and with a high level of concentration. Each movement is also accompanied by deep and concentrated abdominal breathing (Field, 2009).

The effectiveness of yoga in regard to mechanisms for action are believed to involve a number of factors, including enhanced control of the autonomic nervous system, improvements in circulating levels of thalamic gamma-aminobutyric acid, lower levels of circulating cortisol levels, increased vagal activity, and others factors (Culos-Reed et al., 2006; Riley, 2003; Yadav et al., 2012). Even though all of the mechanisms of action are not yet fully understood the evidence is clear that yoga does promote both mental and physical health (e.g., Dhikav et al., 2010; Khalsa et al., 2014; Libby et al., 2012; Williams, et al., 2005). The benefits of yoga in psychological treatment have been well established at this point in time. This particular mind-body therapy will be further examined in upcoming sections of this work in conjunction with more traditional psychological therapies, particularly cognitive-behavioral therapy (CBT) and specific mindfulness-based interventions (e.g., MBCT and MBSR).

Martial arts. The martial arts have long been known for generating within the practitioner in increased sense of confidence, self-regulation, self-efficacy, and self-control. Although there are clear physical benefits to practicing the martial arts (e.g., weight loss, flexibility, blood circulation, lower blood pressure, etc.) there are also many psychological, social, and functional benefits as well (e.g., Chesney et al., 2006; Sun, Buys, & Jayasinghe, 2014). Different martial art forms are being employed daily to help treat a wide range of mental health, behavioral, and social problems and the empirical evidence suggesting clinical efficacy is well supported (Kozasa et al., 2010; Taylor-Piliae, Haskell, Waters, & Froelicher, 2006; Haydicky et al., 2012).

For the purposes of this analytic investigation it will be those martial arts forms that have been

the most widely studied and validated in the literature (specifically in regard to psychological

and behavioral matters) that will be of primary interest. The most thoroughly investigated forms

include tai chi, qigong, karate, and aikido. Other forms, such as jujitsu, judo, and tai kwon do,

are also important to give modest attention to in an analysis of this nature.

Tai chi has been the most widely studied martial art form in regard to mental and physical

health benefits. A review of the literature results in an abundance of empirical investigations that

have been performed to test the effectiveness of this martial art in treating a wide range of issues,

such as insomnia (Kosaza et l., 2010), emotional well-being and stress reduction (Taylor-Piliae

et al., 2006), improving mood states (Chesney et al., 2006), physical functioning and

cardiovascular fitness (Wang, 2011), quality of life and AIDS (Galantino et al., 2005), and a

wide range of additional mental, behavioral and physical health concerns (e.g., smoking

cessation, self-esteem, self-efficacy, anxiety, etc.). Based on the available literature it appears

that tai chi has been the most widely studied martial art across both psychological and medical

fields and the evidence at this time is convincing (e.g., Field, 2012; Wang, 2012).

According to Chesney et al. (2006) tai chi evolved from martial arts and breathing

exercises hundreds of years ago in China. The belief is that practicing tai chi trains not only

the body, but also the mind and *qi* (energy and flow). Tai chi is believed to "quiet" the central

nervous system, improve the elasticity of ligaments, strengthen muscles, calm the mind, improve

wellness, reduce stress, lower anxiety levels, improve sleep, and lower blood pressure, (Chesney

et al., 2006; Galantino et al., 2005; Ko,Tsang, Chan, 2006; Li et al., 2004). In brief, tai chi

means the "grand ultimate fist," which refers to the grand supremacy and uniqueness of this

style. The word "chi" refers to the life force or energy within each individual. Tai chi utilizes internal and external energy in a gentle non-stressful art form that is characterized by slow

controlled movements (Gemmell & Leathem, 2006).

Tai chi is regarded as a low-to-moderate intensity exercise of flowing bodily movements

that is now more widely used to improve mental and physical health (e.g., stress reduction,

ADHD, anger issues, fatigue, traumatic brain injury, etc.) than for actual self-defense (e.g., Field,

2012; Gemmell & Leathem, 2006; Wall, 2005). Tai chi combines both physical and cognitive

techniques and has been demonstrated to be clinically effective across all ages (young and old),

level of fitness/physical ability, and sex. Iaboni and Flint (2013) put forth the merits of

combining cognitive-behavioral therapy with tai chi in the treatment of depression with older

adults. Field (2012) found in her research with children and adolescents that tai chi used in

treatment had a number of mental health and functional benefits when treating those with autism,

ADHD, hyperactive behaviors, anxiety, etc. Gemmel and Leathem (2006) conducted a 6-week

course in tai chi comparing those with traumatic brain injury (TBI) that were assigned the

intervention against a TBI control group. The researchers utilized a within-group design that

tested the immediate effect of tai chi, as well as between-group design testing the long-term

effects of tai chi on TBI and overall functioning. They concluded that the result indicated that tai

chi has significant effects on mood states, emotional functioning, self-esteem, social functioning,

and overall perception of health. Although the short-term effects were much clearer (e.g.,

decreases in sadness, a lessening of anger, tension, and fear, and increases in happiness and

energy), they also stated that more research is needed to support the long-term benefits of tai chi

in treating patients with TBI. At this time, the benefits of including tai chi in psychological and

behavioral treatment is well documented and strongly supported in the literature (e.g., Field,

2012; Sun et al., 2014).

Qigong (also chi gong) is a martial art form that focuses on energy and breath work. Qigong originated in China and is a low-energy expenditure movement therapy that focuses on specific body movements and visualizations that are intended to direct healing mental energies to specific areas of the body. Qigong can be performed standing or in a chair and the practitioner's goal is to create a healing and empowering mind-body experience (Twemlow et al., 2008). Qigong may be especially suited for those of an advanced age or for those that are limited physically as it requires only minor energy expenditure and low-levels of physical strength. Research on the benefits appear to be robust and the evidence seems to indicate that this martial art form is useful across a wide range of issues, such as assisting in the treatment of cardiac conditions, in coping and healing from cancer treatment, in assisting with stress reduction, and helping to improve mood states (e.g., Creamerm, Singh, Hochberg, & Berman, 2000; Pippa et al., 2007; Yeh et al., 2006). Research at Harvard Medical School demonstrated that the mind-body experience of qigong helped patients to improve their mental and physical state when revering from cancer treatment (Kerr, 2002). As with tai chi, the benefits and uses of qigong appear to have clinical utility across a range of psychological and physical health issues.

Karate has also been extensively covered in the literature in regard to its psychological, behavioral, and social benefits. Karate is often regarded as a more aggressive form of martial arts (as compared to styles such as tai chi and qigong). Even though it does emphasize a more physical stance in its methods the evidence contradicts what many think about its effects on behaviors and attitudes. Palermo et al. (2006) argued that karate teaches one self-control, obedience, and compliance. They debunk the myth that karate encourages aggressive behaviors and demonstrated in their research that it actually serves to improve one's self-regulation, executive skills, capacity for concentration, and goal-directed attention. Through their research they were able to highlight how karate increases one's self-efficacy and how it serves to decrease

noncompliance in children with ADHD, reduces antisocial conduct (through addressing

temperamental and cognitive vulnerabilities), and improves the behaviors of children and

adolescents with oppositional defiant disorder (Palermo et al., 2006).

The work put forth by Twenlow et al (2008) supported the use of karate in treating

externalizing and aggressive behaviors. They recommended that karate (as well as other forms

of martial arts) be used with those nonmentalizing individuals that are aggressive (even violent)

that typically do not respond to talk therapies alone. They positioned karate as a useful physical

art form that should be used in combination with traditional talking therapies in the treatment of

those individuals that do not respond well to verbal forms of treatment. They positioned both

karate and jujitsu as effective forms of movement treatments that assist clients in re-tooling their

experiences through kinesthetic experiences that allows for psychological and behavioral healing

(Twemlow, 200). Guthrie (1995) spelled out a very similar experience regarding the value and

power of physical training and healing. In her work she was involved in treating women that

have been victimized physically and sexually. She found that learning karate, which is

essentially a kicking and punching art that involves strengthens one's focus and concentration,

and also involves periods of clam and relaxation, empowered the women and gave them a

newfound sense of psychological mastery and self-control.

Guthrie (1995) argued that talk therapy alone is not enough to treat women that have had

their bodies violated and injured in such ways. She found that karate enabled the women to re-

construct their body images and perceptions of self. Her position was that karate, as a mind-

body healing practice, serves to not only improve women's confidence and self-esteem, but that

it also helps them to become fully empowered individuals that can relearn how to take back control of their own lives. Karate and other forms of martial arts are viewed as complementary forms of therapy. Karate is not viewed as a replacement for traditional talking therapy, but rather it is seen as type of therapy that can help women (or any victim) heal through both the mind and the body (Guthrie, 1995). Other researchers have also found that karate (and other martial arts, such as jujitsu) is effective in lowering aggressive behaviors, improving social relationships and social skills, enhancing attention and concentration, and improving respect for oneself and for others (Kalb & Loeber, 2003; Reynes & Lorant, 2004). As with the other martial arts style karate also appears to have clinical utility across a range of psychological and behavioral conditions.

Other forms of martial arts have also been studied in relation to their abilities to improve psychological, behavioral, physical, and social functioning (such as judo, aikido, and tai kwon do). Lothes et al. (2013) found through their work with aikido that this martial art form can produce positive psychological effects in practitioners. Aikido involves basic body movements that teach the student how to be more aware of their body through controlled balance and breathing. Practitioners of aikido learn to become intimately aware of their bodies, their emotional states, the emotional states of those around them, and how to effectively and safely throw, pin, and effectively adapt with other people (Lothes et al., 2013). The researchers also accredited aikido with also being able to reduce stress, reduce anxiety, instill a sense of inner calm, and increase mindfulness. Twemlow et al. (2008) positioned aikido as an internal softer form of martial arts that is well-suited as a complement to traditional psychotherapy. They stated that this therapeutic harmony resides in the fact that aikido, like psychotherapy, emphasized self-reflection and self-control. Twemlow et al. (2008) also position judo, jujitsu, and tai kwon do as useful movement therapies for treating a wide range of behavioral and psychological issues similar to that of aikido. Although the research on aikido is not as extensive as it is for tai chi and karate, the evidence to date seems to suggest that it also has clinical utility across a range of psychological and behavioral concerns (e.g., treating depression, anxiety, and stress; Shifflett, 1999; Lothes et al., 2013). The martial arts have widely been demonstrated to be an effective and useful complimentary therapy in the treatment of a wide range of psychological and behavioral conditions.

Chapter Seven

Integration of Traditional and Complementary Therapies

The fundamental aim of this work up to this point has been to elaborate on the value and usefulness of employing various CAM therapies into traditional psychological and behavioral treatment. At this point in the work there have been a range of approaches fully examined and connected to the most germane empirical studies. The evidence to date strongly supports the use of CAM therapies in psychological practice. One of the encouraging factors in the use of the various mind-body therapies covered in this analytic inquiry pertains to how accessible these various complementary approaches are to the general population and how accepting the majority seems to have become in recent years (e.g., Kaley-Isley, et al., 2010; Park, 2012). The younger generations also appear to be quite accepting and willing to try adjunctive therapies, which may indicate that CAM approaches may become even more popular over time (e.g., Edenfield & Saeed, 2012). As an example, Pratt (2013) demonstrated in her work with American college students that 100% of those surveyed have used some type of CAM therapy during their lifetime, and 88% reported using at least one or more CAM therapies within the last year. If this is any indication of changing attitudes and acceptance of using CAM therapies then we should expect them to become even more popular with future generations.

This analytic review has put forth a detailed account of a wide range of psychological and behavioral issues that CAM therapies have been shown to be useful in treating. Some of the research presented thus far has demonstrated how beneficial the various mind-body therapies are in treating certain problems, while other studies have used the various mind-body interventions in combination with more traditional psychological treatments (e.g., cognitive-behavioral therapy). As has been demonstrated in earlier sections of this analysis CAM therapies are typically both accessible and practical. The research to date supports the use of CAM therapies in treating a range of issues and presenting problems. Some examples of the psychological issues detailed in this work so far relate to depression and mood states (e.g., Gemmell & Leathem, 2006; Iaboni & Flint, 2013; Wang, 2011), anxiety (e.g., D'Angelo, 2013; West, et al., 2004), stress (e.g., Caldwell, 2010; Kinser, et al., 2013), PTSD (e.g., Jamieson, 2012; Libby, 2012), attention deficit/hyperactivity disorder (e.g., Kalet-Isley, et al., 2010; Krisanaprakornkit, Ngamjarus, Witoonchart, & Piyaavhatkul, 2012), quality of life (e.g., Galantino, et al., 2005; Sun et al., 2014), sleep disorders and insomnia (e.g., Fouladbaksh & Stommel, 2010; Goyal et al., 2014), etc. There has also been a detailed coverage of the benefits and practical use of the highlighted mind-body therapies for a range of behavioral issues across diverse populations, such as externalizing disorders, aggression, and addiction (e.g., Bogels et al., 2008; Palermo et al., 2006; Twemlow et al., 2008).

As can be ascertained from the empirical research examined in this descriptive work, mind-body practices have widely demonstrated their practical value in professional practice. The range of issues and behavioral problems highlighted above does not even cover the entire range of psychological and behavioral issues that CAM therapies have been shown to be useful in treating. Recent research is now demonstrating the effectiveness of certain mind-body practices in treating diseases and mental health illnesses that have not traditionally been investigated or deemed useful for treatment. Vancampfort et al. (2012) conducted a meta-analysis of randomized controlled trials that utilized certain CAM therapies in the treatment and care of those living with schizophrenia. The authors concluded that the evidence to date shows that progressive muscle relaxation, physical exercise, and yoga can all yield positive impacts on mental health outcomes such as state anxiety and psychological distress. They concluded in their meta-analysis that these particular adjunctive therapies have also shown to be effective at reducing both the positive and negative symptoms of schizophrenia and typically result in overall cognitive improvements in the patients (Vancampfort et al., 2012).

CAM Therapies and Physical Disease

While the use of CAM approaches to the treatment of patients with physical diseases is outside the scope of this project, cancer and other physical illnesses certainly can contribute to anxiety, stress, depression, and insomnia (Engelman, 2013; Jacobson & Verret, 2001; Fouladbakhsh & Stommel, 2010; Levine & Balk, 2012; Mehta & Sharma, 2010; Spahn, et al., 2003). A number of empirical research studies have demonstrated the practical use of incorporating certain CAM therapies into palliative and cancer treatment. According to Mansky and Wallerstedt (2006) the use of CAM therapies among cancer patients has now exceeded 50% and the data show that various adjunctive therapies have a range of positive effects on those dealing with the physical and mental problems associated with this disease. The researchers put forth that CAM therapies assist in boosting the immune system in cancer patients, they help to relieve/reduce pain, and they also serve to control the side effects related to the disease and treatment.

In their work, Masky and Wallerstedt (2006) put forth the practical use of acupuncture, hypnosis, massage therapy, and mediation in helping cancer patients to better manage the symptoms associated with cancer and cancer treatment. Mindfulness-based stress reduction was demonstrated to be effective in combination with meditation and yoga in reducing anxiety, insomnia, depression, anger, and stress management (Mansky & Wallerstedt, 2006). DiStasio

(2007) also researched the use of CAM therapies in palliative cancer care and found that mindfulness-based interventions, such as yoga, significantly improve patient's experiences, symptoms, and overall quality of life. The author promoted the integration of mindfulness techniques and yoga into cancer care by presenting the findings on its efficacy in regard to improving sleep, lowering levels of depression and anxiety, and lowering levels of pain. Lee, Pittler, and Ernst (2007) studied the effectiveness of integrating tai chi into cancer treatment. In their work they analyzed several randomized control trials that promoted the benefits of tai chi in regard to relieving cancer-related symptoms and improving overall psychological functioning. As with many of the studies on CAM therapies covered in this work, they also noted a range of methodological flaws and limitations that required improvement over time before any definitive conclusions could be reached regarding clinical efficacy. The bulk of research collected on the management of this particular illness seem to lend support to the position that integrating certain CAM approaches into treatment is therapeutically beneficial in improving patient's physical and psychosocial functioning (e.g., Carson et al., 2007; DiStasio, 2007; Levine & Balk, 2012).

Other researchers have also demonstrated the effectiveness of including certain CAM therapies into the treatment. Smith and Pukall (2009) conducted a meta-analysis of ten randomized controlled trials where women with breast cancer included yoga in their treatment plan. The findings demonstrated, when appropriate, that significant effect sizes were generally noted across the research in regard to the effectiveness of improving psychological well-being across the patients. With this, the researchers also noted some methodological limitations across several of the studies and recommended that further research be conducted in this area.

Kiecolt-Glaser et al. (2010) showed in their research how mindfulness-based interventions can improve cardiovascular functioning, strengthen the endocrine system, and reduce inflammation. They demonstrated that emotional and physical stressors activate various endocrine and immune pathways that promote proinflammatory cytokine production (i.e., certain proteins that promote cardiovascular disease, diabetes, Alzheimer's disease, etc.) and that mindfulness-based interventions (such as yoga and tai chi) can significantly reduce these harmful effects and that it actually has restorative potential (such a lowering cortisol levels and reducing inflammation). Through mixed model and logistic regression analysis they were able to conclude that these particular mindfulness-based interventions can minimize and limit stress-related endocrinological, immunological, and cardiovascular changes that may otherwise prove to be quite harmful to the individual (Kiecolt-Glaser et al., 2010).

Other researchers have demonstrated how various CAM therapies are therapeutically and practically beneficial for neurological and physiological functioning. Daley, Stokes-Lampard, and Macarthur (2009) conducted a meta-analysis of randomized controlled trials that looked at how certain mind-body interventions can help to improve vasomotor and menopausal symptoms across a range of women. They found in their analysis that yoga aerobic exercise has demonstrated significant benefits for managing these particular symptoms and in improving the women's overall quality of life and psychological well-being. The authors also promoted how these particular CAM modalities are also free of harmful side effects and are viable substitutes for pharmacological treatments (Daley, Stokes-Lampard, & Macarthur, 2009). As can be seen from the purview of presented research, CAM therapies are continuing to demonstrating their usefulness and effectiveness across a wide range of psychological, behavioral, and physiological health issues.

Chapter Eight

Cognitive-Behavioral Therapy & the Core Mind-Body Therapies

Cognitive-Behavioral Therapy and Mind-Body Practices

There is a growing body of literature supporting the effectiveness of, and clinical integration of, cognitive-behavioral therapy with certain adjunctive mind-body practices. Chen (2011) explained that the reasons for the growing popularity of integrating various mind-body medicines into clinical practice are due to, outside of the empirical evidence supporting their efficacy, low cost, convenience, and minimal side effects. The researcher noted that yoga tai chi/martial arts, relaxation techniques (e.g., PMR), guided imagery, and meditation are all useful additions to certain therapeutic approaches such as cognitive-behavioral therapy. CBT combined with certain mindfulness techniques has been shown to be very effective in treating a range of psychological and behavioral disorders, such as alcohol and drug addiction, aggression, anxiety, and depression (Adkins et al., 2010; Chen, 2011; Martinsen, 2008). In his research on the benefits of combining CBT with physical exercise, Martinsen (2008) showed how this combination was more effective than any single treatment alone and that it resulted in reduced levels of both anxiety and depression in clinical populations.

In their research on insomnia, Kozasa et al. (2010) positioned CBT as a mind-body intervention and found that CBT was the only approach that resulted in better outcomes than medication. Not only was CBT more effective than yoga, tai chi/martial arts, music therapy, and relaxation techniques, but it also came with none of the harmful side effects commonly associated with drug interventions for sleep disorders. With this stated, the research on this topic also found that the adjunctive mind-body approaches (e.g., yoga, martial arts/tai chi, and relaxation therapy) all resulted in general improvements in sleep time and sleep quality across the patients (Kozasa et al., 2010).

In a randomized control trial investigating symptoms of oral pain, nausea, and vomiting in patients with hematologic malignancies it was found that CBT treatment combined with guided imagery and muscle relaxation resulted in marked improvements across the patients as determined by less mucositis-related pain (Syrjala, Cummings, & Donaldson, 1992). Liossi and Hatira (2003) found in their research that hypnosis and cognitive-behavioral therapy both lowered pain levels, anxiety levels, and behavioral distress among pediatric cancer patients undergoing bone marrow aspiration. Cognitive-behavioral therapy, along with certain mind-body interventions such as yoga, progressive muscle relaxation, and hypnosis, has shown to be an effective integrative approach in treating those undergoing treatment for cancer and other chronic illnesses (e.g., chronic heart-failure, rheumatic diseases), although further research in these areas is needed (Carson, et al., 2007; Mansky & Wallerstedt, 2006; Smith & Pukall, 2009; Sun & Buys, 2014).

Cognitive-Behavioral Therapy and Yoga

At the present time there has not been an ample amount of studies that have combined CBT interventions with yoga, relaxation, and meditation practices in their methodologies but the studies that have been done in this have shown promise (Murphy, 1996; Jones & Johnston, 2000). In a meta-analysis of randomized controlled trials, Hoffman and Smits (2008) found that CBT alone had significant clinical utility in treating participants that were diagnosed with anxiety disorder, but when combined with certain mind-body practices (e.g., meditation) the outcomes were noticeably enhanced across research populations. Khalsa (2004) analyzed more than 150 studies that included yoga as an intervention for the treatment of anxiety in those suffering from a range of medical disorders and found universal improvements. Other research teams have also found that yoga, when utilized as a therapeutic intervention for the treatment of anxiety disorder, resulted in marked improvements in clinical populations (Field, 2011; Kirkwood et al., 2005).

The evidence to date seems to support the stance that using CBT in combination with certain mind-body interventions (e.g., muscle relaxation, yoga, and meditation) is more effective across outcome measures that using any single technique in clinical treatment (Granath et al., 2006). Combining yoga and CBT in clinical interventions with certain psychological disorders appears to show promise based on the most current research. One of the most germane studies located for this particular combination of therapeutic interventions was performed by Khalsa et al. (2014), where they demonstrated that generalized anxiety disorder (GAD) in a treatment-resistance population can effectively be treated with this integrative approach. In their innovative research they developed what is referred to as a yoga-enhanced CBT (Y-CBT) protocol that combines traditional and modified CBT techniques with yoga meditative interventions to help reduce anxiety and depression in a clinical population diagnosed with GAD (N = 32). In the treatment the participants were guided in the Y-CBT protocol which included learning alternate CBT interventions that not only help people to change the content of their thoughts but teach them how to restructure their relations and physiologic interactions with their thoughts (Khalsa et al., 2014).

In their innovative work with Y-CBT therapy Khalsa et al. (2014) conducted 6 week sessions that were focused on affecting change in maladaptive physiologic and cognitive processes in patients with comorbid axis I diagnoses. In their work they sought to reduce dysfunctional thought processes by training the participants to restructure their destructive cognitive and emotional patterns associated with the psychological and bodily symptoms of anxiety. This goal was accomplished by incorporating certain CBT interventions with yoga and meditative-based cognitive solutions that served to reduce ruminative and mind drifting activities of the default mode network in the brain. Through the use of specific self-report measures (e.g., *State Trait Anxiety Inventory* and the *Treatment Outcome Package*) the research team was able to show that Y-CBT has clinical utility in treating GAD and resulted in strong effect sizes across the board (0.97 for the combined treatment of GAD; Khalsa et al., 2014). Although the research had limitations (e.g., an attrition rate of 31%, lack of measures on long-term outcome effects) the findings did show promise in regard to teaching the participants how to better respond to stressful situations and their own thought processes (Khalsa et al., 2014). The findings of this highly relevant study showed great promise in treatment for those suffering from GAD and in reducing state and trait anxiety, depression, panic and overall quality of life. Y-CBT appears to be a very promising new integrative treatment (Khalsa et al., 2014).

The combining of cognitive and behavioral techniques with yoga is still in its infancy. Although there have been some valuable and methodologically sound studies performed with this combination there is still much to be learned about this innovative pairing of approaches. Research has shown how yoga and meditation practices combined with cognitive and behavioral modification techniques can serve to help individuals improve their relations with others by improving their self-management and self-governing skills (Kaliappan, 1998). In this meta-analytic review the researcher put forth how this therapeutic integration helped delinquent boys (N = 35) to significantly reduce their aggressiveness and negative attitudes toward self, family, and society. This was accomplished through four months of behavioral therapy combined with mind-body practices (Kaliappan, 1998).

In another study it was found that social deviance and emotional and behavioral control can be significantly improved by combining cognitive and behavioral techniques with yoga mediation and relaxation techniques (Kaliappan, 1998). This research was conducted for 1 year with 79 murderers serving life sentences. The findings showed overall improvements in areas pertaining to locus of control, decreased risk-taking behavior, improved social skills, improved coping abilities, and marked improvements in emotional regulation and self-control (Kaliappan, 1998).

Although the research to date makes it clear that CBT is a highly effective intervention for the treatment of a range of psychological and behavioral issues it is becoming more evident that using a combination of approaches often result in better outcome measures and clinical improvement (Hoffman & Smits, 2008; Jones & Johnston, 2000). The research conducted by Khalsa et al. (2014) that combined CBT interventions with yoga techniques (Y-CBT) is the most recent and innovative of the integrative approaches and clearly demonstrated how clinically powerful this approach can be. As more evidence comes forth in this area it should become even more pronounced how valuable and effective a mind-body therapeutic combination can be in psychological treatment (Field, 2011; Granath et al., 2006; Kirkwood et al., 2005).

Cognitive-Behavioral Therapy and Martial Arts

The therapeutic integration of cognitive-behavioral therapy with martial arts has been studied across a range of psychological and behavioral issues and has consistently shown promising effects (Huang, Yang, & Liu, 2011; Justina & Man; 2014). Iaboni and Flint (2013) explained how the cognitive restructuring and behavioral changes that comes from CBT training is well complemented by exercise and martial arts. They specifically point to tai chi as an effective integrative modality in the treatment of depression among older individuals. In combination it appears that CBT therapy and martial arts (e.g., tai chi and karate) serve to significantly improve client's motor skills, self-efficacy, sense of control, and overall confidence (Iaboni & Flint, 2013; Palermo et al., 2006). In their research on the positive effects of combining karate with certain behavioral interventions Palermo et al. (2013) found that the efficacy of this type of intervention demonstrated promising results in troubled youths (e.g., oppositional defiant disorder, disobedience, hostility toward authority figures, etc.). They concluded that karate and behavioral interventions leads to significant behavioral improvements and a substantial reduction in problem behaviors. Cognitive improvements were also noted across the sample in relation to goal-oriented attention, executive skills, self-regulation and capacity for concentration (Palermo et al., 2013).

Haydicky et al. (2012) put together one of the most relevant research studies in relation to this particular analytic review. In their research they developed a creative therapeutic approach that combines cognitive-behavioral therapy with mindfulness martial arts (MMA). In their research they recruited 60 adolescent boys with learning disabilities and co-occurring difficulties for a 20-week mindfulness training program. During this program that troubled youths were treated for a range of issues, including externalizing behaviors, oppositional defiant problems, anxiety, ADHD, and conduct problems. The newfound MMA approach integrates specific traditional behavioral techniques, mindfulness meditation, mixed-martial arts, and traditional CBT interventions. They found through their analysis of parent-rated measures that their MMA program resulted in marked improvements relating to all of the aforementioned cognitive and behavioral difficulties across the group of adolescent participants (Haydicky et al., 2012). Similar findings have been found in other studies investigating the benefits of combining martial arts with more traditional psychological interventions (Twemlow et al., 2008).

It was also discovered in the literature that there is a concentrated focus of utilizing CBT interventions combined with certain martial art styles to treat elderly folks for certain issues (Huang et al., 2011; Iaboni & Flint, 2013; Justina & Man; 2014). This is particularly true of the elderly and fear of falling (FOF). In their research with a final sample of 176 community-dwelling residents aged 60 and older, Huang et al. (2011) set out to show how integrating CBT interventions with tai chi martial arts would result in a reduction in the fear of falling, as well as significant improvements in mobility, gait, social support, psychosocial functioning, and overall quality of life. They had three groups in the study (one control group, one CBT group, and one CBT with tai chi group) and found that the CBT-tai chi group displayed the most significant improvements after five months in the program on all measures with the exception of average rate of falls (Huang et al., 2011). This is consistent with the work of Iaboni and Flint (2013) who also investigated CBT interventions, tai chi, depression, and the fear of falling.

Tai chi seems to be positioned as the most appropriate and effective form of martial art (along with qigong) when working with the elderly. One major reason for this is because of the low level of impact and less physical demands placed on the body, compared to other high impact martial arts (e.g., karate, jiu-jitsu, and judo). In another study looking at how CBT can be combined with tai chi to help the elderly reduce their fear of falling and improve their quality of life, Justina and Man (2014) recruited 122 elderly participants (65 years and above) from four elderly community centers for an 8-week training program. CBT was utilized with the aim of increasing self-confidence, sense of control over falling, setting realistic goals, and improving cognitive wellness. Tai chi was utilized with the aim of improving the participant's mobility, balance, gait, self-confidence, physical well-being, concentration, breathing, and to foster a more tranquil mindset (Justina & Man, 2014).

Justina and Man (2014) had two groups for the study. One of the research groups received CBT treatment alone, while the other group received both CBT and tai chi treatments. The findings indicated that both interventions resulted in significant reductions in the fear of falling (but no statistically significant differences were found between the two approaches) and they both improved the participant's levels of confidence and sense of self-efficacy. Mixed effect modeling demonstrated that the combined intervention utilizing both CBT and tai chi did result in significant differences when compared to the CBT alone in regard to self-perceived personal well-being (as measured by the Personal Wellbeing Index/PWI; mean difference: 2.11, 95% CI: 0.20-4.02, $p < 0.03$). The research of Huang et al. (2011) was more optimistic across the range of benefits. In their work they concluded that combining CBT with tai chi resulted in statistically significant effects when compared to control groups and CBT only groups for fear of falling/FOF ($p < .001$), mobility ($p < .001$), social support ($p < 01$), quality of life ($p < .001$). Whereas Justina and Man (2014) only found significant differences between CBT alone and CBT combined with tai chi in the domain of self-perceived well-being, Huang et al. (2011) found significant differences across all outcomes measured (such as FOF, quality of life, improved social support, personal satisfaction, and psychosocial functioning).

One of the reasons for the discrepancy between these research findings may have to do with the noted limitations in the Justina and Man (2014) study. In their work they only allowed relatively healthy elderly individuals to participate in their research. Their inclusion criteria does not allow them to generalize to their findings to more vulnerable populations that are more likely

to fall or that have higher levels of fear of falling. They posited that additional research would

need to be conducted where CBT is combined with tai chi with more frail elderly individuals that

are less mobile and have higher levels of fear in regard to falling (Justina & Man, 2014). This

additional research should help to shed light on any potential additive benefits regarding the

inclusion of tai chi into CBT therapy/training. Huang's et al. (2011) research included a more

diverse aging population so this may explain why they found significant findings across more

outcome measures. The only exclusion criteria in their study was that the participants must not

be terminally ill, must not have unstable health problems, and must not have an artificial leg or

require a leg brace. This was necessary because of the tai chi portion of the intervention where

slow physical movements are required while standing (Huang et al., 2011).

There appears to be a growing body of literature that is looking at the combined effects of

integrating cognitive-behavioral therapy with martial arts. From the research reviewed in this

work it would appear that combining CBT with martial arts (e.g., tai chi) has demonstrated

clinical efficacy across a range of symptoms and issues (Haydicky et al., 2012; Iaboni & Flint,

2013; Twemlow et al., 2008). Some of the most robust findings involved this therapeutic

integration with elderly populations. This is especially pronounced when it comes to the fear of

falling (Huang et al., 2011; Justina & Man, 2014), confidence, self-efficacy, social support,

quality of life (Huang et al., 2011), and perceived well-being (Justina & Man, 2014). Other

studies have demonstrated how traditional psychological interventions (e.g., CBT) when

combined with martial arts can result in social, psychological, and behavioral improvements

across children and adolescent populations (Haydicky et al., 2012; Palermo et al., 2006;

Twemlow et al., 2008). Further research in these areas is needed but at this time the data looks promising.

Chapter Nine

Mindfulness-Based Talk Therapies and Mind-Body Interventions

Mindfulness-Based Interventions and Mind-Body Practices

The empirical evidence that mindfulness-based interventions (MBI's) are useful in treating a wide range of psychological and behavioral problems is now quite strong (Edenfield & Saeed, 2012; Freeman, 2009; Park, 2013). Freeman (2009) examined a range of psychological issues and concluded that MBI's appear to have great potential as a treatment intervention in both clinical and medical settings. Barnett et al. (2014) conducted meta-analyses across a wide range of studies and concluded that MBI's are effective in treating and managing anxiety, stress, depression, hypertension, and pain. Adkins et al. (2010) found in their research that MBI's are effective in reducing aggressive and maladaptive behavior. Bogels et al. (2008) demonstrated in their 8-week training program with a sample of clinically referred adolescents that mindfulness-based training improves attention, lowers impulsivity, and is effective at treating externalizing disorders (e.g., oppositional defiant disorder and conduct disorder). In their research looking at the benefits of teaching mindfulness-based strategies, Caldwell et al. (2010) concluded that developing mindfulness in a population of young adults helped them to improve their quality of sleep, lower their stress levels, improved mood, and improved self-regulation. This is only a small example of the collected information concerning the benefits and use of MBI's in clinical practice and in medical settings. This section of the dissertation will put forth a solid presentation of two of the most popular forms of MBI's, mindfulness-based stress reduction and mindfulness-based cognitive therapy, in connection with complementary mind-body practices.

As presented earlier in this work, mindfulness interventions are rooted in meditative practices (Barnett et al., 2014; Kabat-Zinn, 2003; Park, 2013) and the empirical evidence to date seems to strongly support the benefits and uses of MBI's in combination with CAM therapies across a wide range of psychological and behavioral issues (Field, 2009; Kozasa et al., 2010; Thompson & Gauntlett-Gilbert, 2008). Kabat-Zinn (2003) described mindfulness practices as methods of purposefully paying attention to what is taking place in the current moment in a nonjudgmental way. When mindfulness training is used in combination with certain CAM modalities (e.g., acupuncture, biofeedback, guided imagery, yoga, hypnosis, martial arts, music therapy) the measured outcomes and treatment effectiveness appears to be clinically enhanced (Edenfield & Saeed, 2012; Mansky & Wallerstedt, 2006; Park, 2013).

Baer (2003) conducted a meta-analytic review of 21 studies that utilized mindfulness-based interventions and mind-body therapies (e.g., meditation and yoga) in the treatment of anxiety and depression. It was concluded in his meta-analysis that large effect sizes were indicated across the studies (Cohen's $d = 0.74$, SD $= 0.39$) and that MBI's are clinically effective in bringing those with mild to moderate psychological distress back into the normal range of functioning. It was also noted in his review that more studies needed to be conducted with randomized controlled trials before any decisive conclusions could be reached (Baer, 2003). Park (2013) put forth that mind-body practices are well-suited to mindfulness-based interventions because of how they focus on relations among the mind, body, and behavior. She positioned mediation, yoga, and certain martial arts (e.g., tai chi and qigong) as being particularly well-suited for integration with mindfulness-based interventions for clinical psychologists engaged with behavioral medicine. Others have suggested that additional mind-body therapies are useful for this kind of treatment integration with MBI's, such as progressive muscle relaxation, biofeedback, and physical exercise (Barnet et al, 2014; Field, 2009; Smith, Richardson, Hoffman, & Pilkington, 2005). As can be seen from the presented literature the therapeutic integration of MBI's with mind-body therapies has demonstrated both clinical and practical utility. This will be further demonstrated in the upcoming work focusing on specific therapeutic combinations. The next two section of this work will focus specifically on the primary MBI's and CAM modalities of interest for this analytic review.

Mindfulness-Based Cognitive Therapy and Yoga

Mindfulness-based cognitive therapy (MBCT) represents a modified version of the more traditional mindfulness-based programs. MBCT was originally developed to assist in the treatment of depressive disorders and combines elements of traditional cognitive-behavioral therapy with mindfulness training (Williams et al., 2008; Kabat-Zinn, 2003). Helen and Teasdale (2004) demonstrated how MBCT was a clinically effective way to prevent relapse and recurrence in recovered depressed patients, particularly with three or more previous episodes (e.g., relapse went from 78% to 36% in their clinical sample of 55 patients). MBCT teaches patients to become aware of, and to disengage from, negative and ruminative thinking and to access new modes of thinking characterized by acceptance and the embracement of self. MBCT combines with mindfulness practices (e.g., meditation and yoga) has an additive benefit to traditional care and has shown to be a promising therapy in regard to its clinical effectiveness across a range of psychological and psychiatric conditions (Coelho, Canter, Ernst, 2013; Smith et al., 2005).

As with yoga, MBCT incorporate ideas from Zen teachings about mindfulness, awareness, nonattachment to experience, and the acceptance of experience (Ma & Teasdale, 2004; Levine &

Balk, 2012; Brotto et al., 2009). In sync with certain mindfulness practices such as yoga, MBCT is showing promise in the treatment of a range of disorders beyond depression, such as eating disorders, substance abuse, ADHD, and generalized anxiety disorder (Adkins et al., 2010; Kaley-Isley et al., 2010; Lau & McMain, 2005). Burschka, Keune, Hofstadt-van Oy, Oschmann, and Kuhn (2014) examined research that looked at the benefits of combining MBI's (such as MBCT) with yoga and concluded that the evidence showed that it improved overall quality of life, alleviated depression, and reduced fatigue in clinical patients. Other research has shown that MBCT and yoga are both effective interventions for treating anxiety disorders and that they both take a very similar mindfulness meditative and Zen approach (Evans et al, 2008; Kirkwood et al., 2005; Lau & McMain, 2005). This synergetic approach could be a fundamental reason why this specific combination of interventions can make for such an effective therapeutic integration.

In looking at the benefits of integrating MBCT with yoga, Khalsa et al. (2014) pointed out that both interventions incorporate cognitive and meditative concentration and that they both bring awareness of the physical body (e.g., posture and breathing). The authors noted that including yoga into MBCT can result in improvements in mood, anxiety, emotional well-being, sleep problems/disturbances, hypertension, and headaches. Saeed, Antonacci, and Bloch (2010) suggested that including yoga practices with MBCT can have positive effects as an integrative treatment for depressive and anxiety disorders. They put forth that the therapeutic effectiveness of this integrative approach is superior to no-activity controls and is comparable with established depression and anxiety treatments (e.g., drug interventions such as sertraline and imipramine, and to traditional cognitive-behavioral therapy). The integration of complementary and alternative therapies has shown great promise in research and clinical practice and therapeutic integration is only going to continue to increase over time (Hoffman et al., 2010; Saeed et al., 2010; Teychenne, Ball, & Salmon, 2008).

Although there still needs to be much more research done in relation to the therapeutic effects and increased benefits of integrating yoga with MBCT the evidence to date seems to suggest that it is a valuable combination. Khalsa et al. (2014) argued that combining yoga with mindfulness cognitive therapy is a very effective clinical intervention across a range of psychological issues (e.g., depression, sleep problems, state and trait anxiety, and panic disorders). Saeed et al. (2010) point out that many of the studies in the literature in this area show multiple methodological weaknesses and that further research is needed in certain integrative areas. Lengacher (2012) agreed with this position and noted that there is a paucity of research to date that looks at the combined effects of yoga and mindfulness-based cognitive therapy. Although Lengacher stated that the evidence to date is suggestive that combining yoga with MBCT looks promising there still needs to be more rigorous research to verify its efficacy. Khalsa et al. (2014) were firm in their stance that combining yoga with MBCT and CBT is therapeutically effective and that these can be a very useful integrative approaches for treating a range of psychological conditions.

Mindfulness-Based Cognitive Therapy and Martial Arts

At this point in time the research investigating the specific integration of mindfulness-based cognitive therapy and martial arts is sparse. After an extensive literature review was conducted across a handful of databases only a few studies were identified that were connected to this particular therapeutic combination or some derivation of it. Most of the work in this area is connected to research looking at the benefits of combining mindfulness training (such as

MBCT and MBSR) with martial arts in the treatment of patients with multiple sclerosis (Mills & Allen, 2000; Grossman et al., 2010; Simpson et al., 2014). Mills and Allen (2000) performed a randomly controlled trial in their investigation of the effects of MBI's and tai chi and reported significant findings in improvements in depression ($p < 0.01$) and on certain physical measures (e.g., standing balance; $p < 0.05$). They did not find statistically significant results in regard to anxiety or pain reduction. The findings also indicated that those receiving the mindfulness training along with the tai chi training showed improvements across certain psychosocial outcomes, such as quality of life and general symptom improvement (Mills & Allen, 2000).

In a 20-week mindfulness training program, Haydicky (2010) examined the effectiveness of mindfulness martial arts (MMA) in the treatment of 65 boys aged 12 to 18 with learning disabilities. Mindfulness martial arts is a therapeutic combination of mindfulness-based meditation and cognitive-behavioral therapy, essentially a derivative of MBCT. The martial arts aspect of this approach is known as "mixed" martial arts and integrates a variety of techniques from various disciplines (e.g., karate, judo, jiu-jitsu, and tae kwon do). The findings of the study showed that this particular combination of mind-body interventions resulted in significant improvements in the treatment group (compared to the control group) across all measures, including externalizing behaviors, conduct problems, oppositional defiant problems, anxiety, and ADHD. MMA, which is a combination of mindfulness meditation, martial arts, and cognitive therapy, demonstrated promise as a treatment option for youth with learning disabilities and the highlighted co-occurring diagnoses (Haydicky, 2010).

Simpson et al. (2014) argued that mindfulness based interventions such as MBCT operates differently from traditional CBT because of its strong emphasis on self-awareness and regular self-practice. In their review of the research on utilizing a combination of MBI's (such as

MBCT) and tai chi in the treatment of those with multiple sclerosis they found that (although the research is still scant in this area) the evidence looked promising and that combining mindfulness-based interventions with mind-body practices appears to be useful in improving patients mental health (e.g., depression), physical health/abilities (e.g., movement, pain, and balance), and psychosocial functioning (e.g., personal well-being and social participation; Simpson et al., 2014). Additional research in this area has also shown promise in regard to the use of mindfulness-based training in the treatment and improvement of quality of life and well-being for those managing life with multiple sclerosis and other chronic disorders (Grossman et al., 2010; Tavee, Rensel, Pope, Planchon, & Stone, 2011).

Other research teams have also assessed the standing literature in regard to the integrative potential of mindfulness practices and mind-body therapies. Simkin and Black (2014) performed a review of the literature and concluded that the therapeutic integration of mindfulness techniques, such as MBCT and MBSR, with body-mind techniques, such as tai chi and yoga, have been shown to produce enduring neurobiological changes in the brain and physiologic improvements in body function for those that continue to practice the prescribed techniques. The research team determined that these findings hold across a range of problems, including depression, anxiety, somatic functioning, and other behavioral and emotional symptoms (Simkin & Black, 2014). Some researchers are not as optimistic about this and stated that more research is needed at this time as the evidence from rigorous studies and randomly controlled trials is still scant (e.g., Saeed et al., 2010). Burschka et al. (2014) stated that the use of tai chi in combination with mindfulness-based interventions to treat certain clinical populations (e.g., patients with multiple sclerosis and other neurologic diseases) holds therapeutic potential but further research is needed across larger samples to determine the underlying working mechanisms.

At this point in time there is empirical evidence that suggests that MBCT in combination with martial arts is effective in treating a range of mental and behavioral issues (Haydicky, 2010; Simkin & Black, 2014; Simpson et al., 2014) but more research is clearly needed in this specific area. The use of MBI's that incorporate cognitive therapies with martial arts, particularly tai chi and qigong, have shown potential across certain studies but more controlled research is needed before any conclusive statements can be made (Grossman et al., 2010; Tavee et al., 2011; Mills & Allen, 2000). With this stated, the empirical data to date consistently leans in the suggestive direction that combining certain mindfulness-based interventions such as MBCT with certain mind-body practices such as tai chi, qigong, and yoga, has a therapeutic potential across a range of psychological and behavioral issues (Burschka et al., 2014; Simkin & Black, 2014; Simpson et al., 2014).

Mindfulness-Based Stress Reduction and Yoga

One of the main components of mindfulness-based stress reduction (MBSR) is yoga, which includes a combination of physical postures, meditation, and breathing techniques. This combination of practices has been shown to improve coordination, flexibility, concentration, sleep, digestion, reduce stress, and lower blood pressure (Lin et al., 2011). The most common form of yoga that is integrated with MBSR is Hatha yoga, which focuses on mindful movement, concentration, and awareness (Lengacher, 2012; Park, 2013; Simpson et al., 2014). According to Smith et al. (2005) MBSR combined with Hatha yoga exercises is traditionally delivered in a group setting across 8-weeks of classes (2.5 hours per class). Other gentle yoga techniques have also been studied in combination with MBSR and the results seem to indicate equivalent outcomes (Carlson et al., 2004; Grossman et al., 2010). For the purposes of this work any form of yoga practiced with MBSR was taken as relevant and in line with the scope of this descriptive analysis. Because yoga is central to the MBSR research covered in this analytic work the designated MBSR-y will be used across this section to avoid descriptive redundancy and to ensure that it is clear that all reviewed research used this particular therapeutic integration.

A good portion of the relevant research discovered for this work revolved around palliative care for cancer patients. There appears to be a good deal of evidence at this time that MBSR-y has a range of positive effects for those battling various forms of this deadly disease (Lengacher, 2012; Lin et l., 2011; Mansky & Wallerstedt, 2006; Smith et al., 2005). Lin et al. (2011) conducted a meta-analytic review of randomized controlled trials to discern therapeutic efficacy on the quality of life, psychological health, and physical health of cancer patients (10 RCTs, total N = 762). The authors found generally consistent results regarding therapeutic effectiveness in treating distress, stress, anxiety, depression, quality of life, and physical health. Mansky and Wallerstedt (2006) studied palliative care and symptom management in cancer patients and concluded that MBSR-y has demonstrated therapeutic potency in decreasing anxiety, mood disturbance, depression, anger, and stress.

Smith et al. (2005) completed a systematic review and critical appraisal of both randomized and nonrandomized controlled studies to see if MBSR-y had clinical efficacy in supportive care for cancer patients. Although they were critical of some of the studies performed and the reporting of the specific interventions used (e.g., types of yoga included in the studies, content of weekly sessions) they still concluded that this type of supportive intervention is generally safe and promotes psychological and physical well-being (e.g., reduced stress, improved mood, improve sleep, improved quality of life, less pain, and reduced anxiety). With this positioned in the final analysis, the authors still made it clear that further research and more controlled outcome measures are still needed (Smith et al., 2005). Other researchers have also taken the position that MBSR-y is an effective way to improve cancer survivor's lives and the lives of current patients through chronic disease management and the teachings of meditative techniques and body awareness (Lengacher, 2012; Simpson et al., 2014).

Further research studies have also shown that MBSR-y has clinical efficacy in the management and treatment of cancer patients (Carlson et al., 2003; Carlson et al., 2004). In their work looking at the benefits of incorporating MBSR-y into treating patients with breast cancer and prostate cancer, Carlson et al. (2003) concluded that this intervention assisted in patient improvements across several domains, including symptoms of stress, sleep quality, and overall quality of life. In their research that spanned 8 weeks, and that recruited 49 breast cancer patients and 10 prostate cancer patients, they demonstrated that the patients immunity profiles (as measured by cytotoxic cells, lymphocyte counts, T cell production, and cytokine production) significantly improved after MBSR-y intervention. They argued that these outcomes are consistent with shifts in depressive symptoms and found that the gentle yoga, meditation, and relaxation techniques helped to get patients back into normal immune profiles combined with improved psychological functioning (Carlson et al., 2003). Utilizing the same MBSR-y intervention, Carlson et al. (2004) investigated the therapeutic effects of this 8-week intervention on 59 patients (breast and prostate cancer). They found that the MBSR-y intervention resulted in significant improvements in quality of life, decreased stress symptoms, improved sleep, and improved mood. The study data indicated that changes in hypothalamic-pituitary-adrenal axis functioning, salivary melatonin levels, salivary cortisol levels, and dehydroepiandrosterone-sulfate correlated significantly with patient outcomes (Carlson et al., 2004). According to the researchers these physiological changes were the direct result of the MBSR-y treatment intervention and further demonstrated its clinical efficacy in improving the lives of cancer patients.

Multiple sclerosis is another disease that has been well studied in relation to MBSR-y interventions. Grossman et al. (2010) recruited 150 participants to assess the efficacy of mindfulness-based interventions for improving the lives of those living with multiple sclerosis. The research team assigned 76 participants to the intervention group and 74 participants to the control group so that they could make valid outcome assessments after the 8-week program. They concluded that the data from the study provided Class III evidence that this mindfulness-based intervention improved fatigue, depression, and quality of life across the members of the experimental group ($p < 0.002$, effect sizes $= 0.4 - 0.9$). They concluded that health-related quality of life can be significantly improved for those living with multiple sclerosis and also positioned MBSR-y interventions as a having treatment implications for those living with other chronic illnesses (Grossman et al., 2010). Burschka (2014) supported this stance and put forth that physical and mindfulness interventions, such as MBSR-y, have shown promise when it comes to treating multiple sclerosis patients. Although the authors noted that the effort in this area still needs more structured research the findings to date seem to suggest clinical utility.

In a systematic review of the literature, Simpson et al. (2014) found that the evidence at this time seemed to support the position that certain mindfulness-based interventions such as MBSR-y do improve the lives of those living with multiple sclerosis by alleviating pain, stress, anxiety and depressive symptoms. The authors noted that MBSR-y appears to have positive neuroplastic, neuroendocrine, and immunological effects that can have the result of improving both mental health outcomes and physical outcomes for patients. Although they were quite optimistic in their position on the benefits of MBSR-y treatment for those with debilitating and chronic diseases, such as multiple sclerosis, they were still clear that further studies were required to clarify best methods and outcome measures (Simpson et al., 2014).

White (2012) found in her study that MBSR-y was effective in reducing psychological and physical problems among a population of school-aged girls. Female students from two public elementary schools were randomly assigned to intervention and waitlist groups. After the 8-week intervention the group of girls that received the MBSR-y intervention showed marked improvements in self-regulation, self-esteem, coping abilities ($p < 0.05$), and stress levels ($p < 0.01$) compared to the control group. Although there is clearly a need for more research looking at the efficacy of mindfulness training and awareness-based practices among child and teenage populations the findings from the study seems to suggest that MBSR-y interventions have utility when it comes to reducing stress, improving emotional development, and improving social skills and coping skills (White, 2012). At this time it appears that MBSR-y has demonstrated promise as an effective intervention for treating a range of psychological and physical issues across diverse populations (Barnett et al., 2014; Burschka et al., 2014; Lin et al., 2011; Park, 2013; Simpson et al., 2014; White, 2012).

Mindfulness-Based Stress Reduction and Martial Arts

Mindfulness-based stress reduction is the most well researched mindfulness-based intervention in and has been widely shown to be effective in treating a range of psychological issues, such as chronic pain, stress, depression, and anxiety (Simpson et al., 2014). When used in combination with certain martial art forms MBSR can be even more effective in alleviating certain psychological issues and behavioral problems (Lothes et al., 2013; Wall, 2005).

Mindfulness practices can be integrated with martial arts to help individuals learn how to self-regulate their feelings and the triggers that arouse feelings, how to be more aware of their mind and body as an organic whole, how to contain emotion-laden stress, and how to promote personal competence and cognitive control (Burschka et al., 2014; Ott, 2002).

In their work demonstrating the positive benefits of mindfulness practices on overall mental health, Lothes et al. (2013) hypothesized that aikido training combined with mindfulness programs such as MBSR would result in significant increases in mindfulness scores, heightened awareness and improved psychological functioning. Aikido teaches practitioners to be more aware of one's own body and the bodies of others, to be more aware of one's emotional state and the emotional states of others, and to improve coordination, breathing, and balance. With these specific approaches, goals, and benefits it is believed that aikido is a natural fit with mindfulness-based techniques such as MBSR. Through measurements on the Mindfulness Attention Awareness Scale and the Kentucky Inventory of Mindfulness Skills the research team was able to show from their longitudinal study of 20 aikido students that the program resulted in statistically significant improvements across all measures (Lothes et al., 2013). Similar significant findings have also been observed with other martial arts forms (e.g., karate, tae kwon do and mixed-martial arts) in relation to increased mindfulness, self-regulation, self-control, and overall psychological improvement (Haydicky et al., 2012; Twemlow & Sacco, 1998; Twemlow et al., 2008).

In a clinical project with 6th-grade girls and 8th-grade boys in a Boston inner-city middle school, Wall (2005) demonstrated that combining MBSR with tai chi can have benefits across a range of measures, including improved relaxation, improved overall well-being, less reactivity, increased self-care, and a more keen sense of interconnectedness. During the five week course the students were taught to improve their concentration and awareness and to calm both the mind and the body. It was proposed that integrating MBSR with tai chi not only served to enhance the overall outcomes (compared to any single technique alone) but it also taught them better monitor and contain their own aggressive impulses. Combining martial arts (e.g., tai chi and qigong) with MBSR interventions was shown to be an effective modality for middle school students and was recommended as a productive transformational tool for educational programs and inpatient and outpatient psychiatric settings for pre-teens and teenagers, and for use in pediatric settings (Wall, 2005). Twemlow et al. (2008) made similar recommendations for integrating martial arts practices with psychological interventions to contain certain harmful behaviors and to improve overall functioning.

In their research on the benefits of combining tai chi with MBSR among a mixed group of 17 patients with multiple sclerosis and patients with peripheral neuropathy, Tavee et al. (2011) concluded that this integrative intervention was successful in reducing both fatigue ($p = 0.035$, Modified Fatigue Impact Scale) and pain ($p = 0.031$, Visual Analogue Scale). In this 8-week nonrandomized control trial the researchers had a Buddhist monk teach the patients walking meditation, mindful breathing, and mindful movement, which included both tai chi and qigong techniques. The Buddhist monk taught the participants breath and posture awareness, while teaching them how to cultivate compassionate feelings toward themselves. This goal was accomplished through the mindful movements of qigong and tai chi (Tavee et al., 2011). Mills and Allen (2000) found similar results in their research with patients with multiple sclerosis ($N = 16$). In their randomized controlled trial that utilized mindfulness breathing and mindfulness movement they determined that patients receiving the intervention (i.e., tai chi/qigong and MBSR) showed significant improvements as measured by the Profile of Mood States ($p < 0.01$)

and in physical outcomes (e.g., single leg standing balance). Effect size was still significant at the 3 month follow up ($p < 0.05$). Patients did not show significant improvements for anxiety or fatigue as measured by the Profile of Mood States ($p > 0.05$; Mills & Allen, 2000).

In their research with mildly disabled multiple sclerosis patients, Burschka et al. (2014) found that integrating mindfulness-based interventions, such as MBSR, with tai chi can alleviate suffering from depression, mobility impairment, and fatigue. The study compared the effects of the MBSR-tai chi treatment with a group of 15 patients to a "treatment as usual group" (control group) of 17 patients. Depression was significantly improved after the intervention in the treatment group but remained stable in the control group and demonstrated a significant main effect [$F (1,27) = 6.61$, $p < 0.05$; as measured by the Center for Epidemiological Studies Depression Scale]. Life satisfaction in the treatment group was significantly improved [$F (1,24) = 8.64$, $p < 0.01$; as measured by the Questionnaire of Life Satisfaction], as was fatigue [$F (1,25) = 7.83$, $p = 0.01$; as measured by the Fatigue Scale of Motor and Cognitive Functions]. Measures on balance and coordination also showed significant improvements across all measures [$F (1,30) = 5.70$, $p < 0.05$, and $F (1,30) = 4.89$, $p < 0.05$, respectively]. All measures across the control group remained stable and did not improve across any of the outcomes studies in this research which strongly suggests that integrating MBSR with tai chi/qigong has significant effects across multiple measures and improves patients overall quality of life (Burschka et al., 2014).

Some of the methodological problems across the studies looking at MBSR and martial arts integration with multiple sclerosis patients include relatively small samples sizes and participant attrition. Mills and Allen (2000) had an attrition rate of 12.5%, Burschka et al. (2014) had an attrition rate of 16%, and Tavee et al. (2011) had an attrition rate of 43%. Difficulty with mobility, pain, transportation issues, bereavement, and dislike of the intervention were the most cited reasons for dropping out. Either way, the findings to date seem to demonstrate that combining MBSR with martial arts (particularly tai chi and qigong) has the end result of improving the lives of those living with multiple sclerosis and peripheral neuropathic diseases (Simpson et al., 2014; Tavee et al., 2011).

In the final analysis it seems clear that mindfulness-based techniques combined with martial arts has shown to be an effective therapeutic intervention across a range of populations and psychological, behavioral, and physical conditions (Burschka et al., 2014; Mills & Allen; 2000; Tavee et al., 2011; Wall, 2005). Aikido, tai chi, qigong, karate, tae kwon do, jiu-jitsu, and mixed-martial arts have been demonstrated to be beneficial complements to mindfulness-based interventions such as MBSR and to improve psychological and behavioral functioning across a range of issues (Haydicky et al., 2012; Lothes et al., 2013; Simpson et al., 2014; Twemlow et al., 2008; Wall ,2005). Additional research is needed to further verify these preliminary findings but the data at this time seems to strongly indicate both clinical and practical utility. Further research will help to determine what the exact underlying working mechanisms are and how they can best be applied to service populations in need (Burschka et al., 2014; Simpson et al., 2014; Wall, 2005).

Chapter Ten

Evidence-Based Demonstrations

Thus far in this analytic presentation I have focused largely on empirical research and have brought forth a copious amount of data and findings from both randomized and nonrandomized studies. I believe that this presentation has made a solid case for the effectiveness and usefulness of integrating CAM therapies into clinical practice. Combining more traditional talk therapies (e.g., CBT, MBSR, MBCT) with the aforesaid mind-body modalities has been shown to generally result in enhanced positive clinical outcomes across a wide range of presenting issues. The previously discussed findings are of significant importance to the psychotherapeutic community at large and my overarching desire is to get this valuable information disseminated to interested practitioners so that they can be more effective in treating their client's needs.

As put forth at the beginning of this exposition, the use of CAM therapies is increasing across the United States. It appears that increasing numbers of people are becoming more willing to give these complementary treatments a chance and are looking for healthy and safe adjuncts to traditional treatments. This growing acceptance can be seen in clinical practice today. Psychologists and psychotherapists across the nation are now including CAM therapies in their practice and the results seem to be largely in line with what has been presented in this work. The psychotherapeutic community would be served well by learning more about how to effectively integrate these useful CAM modalities with their traditional therapeutic approaches. In a final presentation I would like to bring forth a number of case examples that demonstrate how integrating mind-body therapies helps to improve client's lives and serves to improve clinical outcomes. The case examples selected will highlight and revolve around the primary CAM therapies examined in this presentation – yoga and martial arts.

Psychotherapeutic Relevance: Clinical Vignettes

Case Vignette: Samantha

Samantha was a 16 year-old student of mixed African-American and Caucasian ancestry who was residing at a residential treatment facility in the Northeastern United States. When she was admitted into the treatment facility Samantha was diagnosed with having significant social phobias, generalized anxiety, mild mental retardation, post-traumatic stress disorder, and psychotic disorder NOS. Prior to entering the residential facility Samantha was living under foster care after being removed from her highly abusive biological parents. She was removed from the home around 7 years prior to entering residential care after many reports of her being abused and neglected by her biological parents. For example, she would often come to school with an unkempt presentation and smelling of foul odors. Her foster parents had her placed into the residential facility after they realized that they could not keep her safe and could not manage her presenting problems (Spinazzola, Rhodes, Emerson, Earle, & Monroe, 2011).

After being admitted into the treatment facility she began to exhibit abrupt shifts in mood. She would have episodes of extreme irritation, crying, and explosive self-harming and aggressive behaviors. She would also engage in binge eating and hoarding of food, which would typically be followed by periods of shame, withdrawal, and dysphoria. Samantha had an intense feeling of being interpersonally rejected and that she was inferior to others and was always being treated unfairly. She had difficulty communicating and getting along with others and was not adept at reading social cues. She was also fixated on negative self-attributions, such as referring to herself as "fat" and an "ugly klutz." Samantha struggled with severe emotional and behavioral dysregulation (Spinazzola et al., 2011).

During her time at the residential program Samantha received extensive individual and group therapy. Group therapy focused on improving coping skills and social skills. Individual treatment included behavioral and cognitive approaches that complemented the group therapy by furthering her repertoire of coping skills and social skills. Individual therapy also focused a good deal on working through the trauma that she had to endure when she was a child. The treating therapists determined that Samantha should also engage in recreational activities to help her with her emotional regulation and to improve her overall sense of self (Spinazzola et al., 2011).

After stabilizing from an acute presentation of psychotic features for which she was hospitalized for one month, Samantha was returned to the residential facility at which time she requested that she be allowed to participate in Hatha yoga programming. Her treatment team thought that this would be a good practice for her due to the structured nature of yoga. The treatment team also felt that the yoga would help Samantha with self-esteem and body issues, help her to recognize signals of anxiety, and help her to learn ways to use breathing and mindfulness practices to increase self-control and to prevent escalation of distress. The team felt that learning yoga would assist her in controlling her maladaptive outbursts that often resulted in harmful consequences to self and to others (Spinazzola et al., 2011).

The yoga sessions took place two to three times each week and were intentionally kept brief (usually around 15 minutes) to avoid feelings of being overwhelmed and to create a gradual familiarity with somatic arousal and mastery of routines. There were three different configurations used in the sessions. One configuration was classroom-based and included the yoga teacher and Samantha's peers. The second configuration was a triad that included Samantha, the yoga instructor, and a residential staff member. The third configuration consisted of four group members and included Samantha, one nonconflictual peer, a residential staff member, and the yoga instructor (Spinazzola et al., 2011). The sessions were a combination of chair-based practices and mat-based practices. For the first three months of yoga training Samantha was very compliant and attending her sessions regularly. During this time it was observed by facility staff that her number of behavioral incidents had been reduced to almost half of what they were prior to the yoga training. Samantha enjoyed all three yoga configurations and she identified distinct benefits from each of them. For example, in the classroom-based configuration she was able to take on a mentoring role to new peer. In the triad configuration she was able to attend more to her own body sensations, improve her strength and flexibility, and increase her focus (Spinazzola et al., 2011).

Through the practice of yoga, in combination with group and individual therapy, Samantha significantly improved her somatic awareness and her responsiveness to her bodily needs. These improvements carried over into other important areas of her daily life and helped her to alleviate her feelings of anger, sadness, and loneliness. She became more cooperative with staff and her interpersonal skills and relations dramatically improved. Residential staff observed that she was also better able to regulate her eating behaviors and was much better at verbalizing when she felt full. Her hoarding of food ceased and she was no longer binging like she had been prior to the yoga treatment. After six months of yoga treatment Samantha reported that she was more aware of her feelings and that she was better able to monitor her behaviors. She stated that yoga helps her "tune in" and helps her when she is "too amped up and needs to chill out or eat something or talk something out with a staff before I totally lose it." She was also happy in the physical improvement that she received from her yoga training. She pointed out that she felt better about her body ("I don't hate on my body like I used to"), that she was stronger, and that she was "less flabby" than before (Spinazzola et al., 2011).

After nine months of yoga therapy Samantha had demonstrated a clear deepening level of personal investment in and ownership of this self-regulating practice. She gained a clear sense of personal empowerment through yoga training and mindfulness practices. Samantha's life had improved dramatically during her nine months of yoga therapy (coupled with group and individual therapy). "Clinical observations of Samantha's continued gains in affect regulation and impulse control, seem indicative of a burgeoning ability to listen to her body and respond appropriately to its signaling of needs such as hunger, fatigue, anger, and distress" (Spinazzola et al., 2011, p. 436). One of the primary goals of trauma-sensitive yoga is to assist students in reconnecting with, and to feel safe in, their bodies. This vignette demonstrates a successful case where incorporating yoga training into a congregate care setting resulted in significant emotional and behavioral improvements in this particular young female.

Case Vignette: Danny

Danny is a Caucasian male who was enrolled in residential school as a response to being expelled from school for truancy, fighting, and assaulting a male teacher. At the time he was enrolled in the residential school he was 17 years of age, 6 feet tall, and around 190 pounds. He would spend hours each day lifting heavy weights and had built a very muscular and strong ("dangerous") body. Several years prior to being admitted into the residential school Danny's mother had committed suicide. His biological father was severely abusive toward him and reports from social services stated that as Danny aged he began to seek out violent retaliations against his estranged father with increasingly assaultive behavior (Spinazzola et al., 2011).

Danny was diagnosed by his clinical team as presenting with conduct disorder and intermittent explosive disorder. At the time of admission to the treatment facility Danny lacked cause-and-effect thinking, had an exaggerated startled response, and displayed difficulties comprehending social rules, social cues from others (Spinazzola et al., 2011). He had difficulty developing healthy relationships with others and was well known for physical posturing and making verbal threats to staff and peers. He had a chronic tendency to interpret interactions with other males (staff and peers) as being aggressive and would often respond in kind. It was common for clinical and residential staff to complain to the program director or their immediate supervisors about Danny's aggressiveness and verbally abusive behavior. A number of staff made requests that Danny be reassigned to a different classroom, clinician, or floor in the residence. Numerous staff made it clear that they did not want to have primary responsibility for him (Spinazzola et al., 2011).

During his second month at the residential treatment facility Danny was evaluated as having little awareness of his somatic states of arousal and was out of touch with his emotions and their relationship to his dysregulation. Danny was referred to yoga by his therapist who believed that what he needed was to feel safe in his body. His therapist felt that he needed something physical to do that was not violent, dangerous, or frightening (Spinazzola et al., 2011). At this point in time it was deemed that Danny was a good candidate for a Hatha yoga program. It was believed that trauma-informed yoga therapy within the context of a quad configuration was the best structure for treatment. The quad consisted of Danny, the yoga instructor, a staff member, and one peer. The chosen staff member was athletic and one of the few workers that generally got along with Danny. The chosen peer was younger, small in stature (non-threatening), and was known for his sense of humor, which Danny seemed to respond to (Spinazzola et al., 2011).

During the first yoga session Danny was very tense and uneasy. He paced back and forth in the room and was hyperventilating and breathing in an irregular and rapid manner. His first words were, "I am so angry, I don't know what to do" (Spinazzola et al., 2011. p. 339). With this the yoga instructor realized that Danny needed to do something with his body, get in touch with his emotions, and control his breathing. Danny agreed to the training but was reluctant and expressed that he did not think that it would work for him. The yoga instructor began the session with breathing techniques. The goal here was to get Danny in touch with his body, his breathing, and the "physicality of breath." After some time the group began incorporating neck and shoulder rolls. The breathing and movement exercises continued on for 30 minutes and allowed Danny an opportunity for positive modeling, interpersonal attunement, and a sense of normalcy (Spinazzola et al., 2011).

Danny quickly took to the yoga training and appeared to really enjoy the various warrior positions and the deep breathing exercises. Several residential staff members, most of whom had a tumultuous relationship with Danny, voiced how much Danny had changed since his yoga training had begun. He appeared to be visibly less tense after the yoga sessions. His face was more relaxed and his physical posturing changed dramatically. His breathing was more controlled and relaxed.

After several weeks of yoga training Danny exhibited marked improvements across several domains. The yoga instructor realized that Danny had developed a much deeper awareness of his body and encouraged Danny to explore this further in his within the context of his individual therapy sessions. Danny agreed and finally opened up in his therapy sessions about his anger toward his father and his desire for retribution (Spinazzola et al., 2011). This newfound ability and willingness to open up about his feelings enhanced the therapeutic relationships and allowed for significant gains in the therapy sessions:

> Clinical supervision of his psychotherapist focused on helping his therapist manage her own fear and anxiety-based emotional and physiological responses to these often disarming verbalizations. This enabled Danny's therapist to maintain a nonjudgmental stance of openness and physical groundedness where Danny's impulses could be normalized within the framework of psychoeducation about the impact of exposure to violence and trauma on the human stress response. This approach gradually fostered in Danny a sense of connection with his therapist, achievement of which was a critical precursor to willingness to consider and practice prosocial coping and self-regulation skills. In turn, Danny's therapist was able over time to redirect these conversations to discussions of social justice, in which Danny showed great interest. Eventually, and in tandem with the increased somatic awareness he was developing in his yoga sessions, Danny and his therapist were able to explore the notion of what it meant to feel 'strong on the inside' and to entertain alternatives to violence in response to the experience of being wronged by another. (Spinazzola et al., 2011, p. 440)

This case study provides an additional example of how trauma-sensitive yoga can be a beneficial complement to traditional talk therapy. In this case example we can see how this young man was able to become more aware of his physical body and somatic sensations, how to control his breathing, and how to better regulate both his behaviors and emotions. The Hatha yoga training proved to be a quite valuable addition to his overall therapeutic treatment and helped him to better communicate with his therapist, which in turn fostered a much safer and healthy

therapeutic relationship between the two.

Case Vignette: "Jessie"

Jessie had a history of aggressive and violent behavior. He was known to mistreat his peers

in a rather affectless manner and also caused physically injuries to others. Jessie was ordered by

the court to receive treatment for his sadistic bullying and violent behavior. Even though his

parents were reluctant to place Jessie into therapy they complied with the court order (Twemlow

et al., 2008). Jessie's family background consisted of two abusive and violent parents. His

father was a severe alcoholic and was violently abusive toward him. His mother also exposed

him to severe domestic violence and was detached from him in an emotional way. His parents

divorced when he was eight years of age and his mother's subsequent boyfriends also physically

abused him (Twemlow et al., 2008).

Jessie had developed poor social skills and had difficulty interpreting the social cues of

others. He was known for having bad manners and seemed to develop a pathological

identification with his violent father. Jessie stated that his father had taught him to "whack"

anyone that ever crossed him. Jessie was constantly humiliated and harassed at school by his

peers. After puberty Jessie seemed to develop a sadistic stance toward others. Jessie

displayed many of the symptoms of a traumatized individual. He was frightened of being seen

as "soft" or gentle and he would in turn take a very macho and grandiose stance toward others.

He seemed to relish in demeaning and humiliating others (Twemlow et al., 2008).

After a period of time in psychotherapy his therapist decided that martial arts training

would be a good complement to his overall treatment plan and would help him to better control his behaviors and his emotions. His therapist regularly attended the martial arts classes and used them as opportunities to work on the concepts that had been discussed during their therapy sessions (Twemlow et al., 2008). During the classes Jessie would frequently act in an overly aggressive manner and few students liked to spar with him. He even acted very aggressively toward the female students and was often required to publicly apologize to them and to perform services as a part of his reparation. Jessie had a problem with overestimating the aggressiveness and intentions of others and seemed to view these acts as threats toward him. Over time his behaviors and hostile attributions toward others improved significantly and after five years of intensive training he eventually received a black belt (Twemlow et al., 2008).

As a part of his black belt training Jessie was required to perform community service. He elected to teach self-defense training to severely handicapped children in a local institution. These children were confined to wheelchairs and many of them could not speak and had brain damage. Over time Jessie developed a deep sensitivity for these children and became intimately attuned to their limited physical and verbal skills. Jessie would frequently talk to his therapist about his interactions with these handicapped children and how they constantly reminded him of his abusive father and how he always feared being injured or brain damaged by these attacks. These "flashback" discussions proved to be very therapeutic for him and for his overall healing (Twemlow et al., 2008). Through developing these meaningful relationships with the children, and through being guided by his attuned therapist and martial arts instructor, Jessie was able to ameliorate his "toxically spoiled alien self." During the treatment and martial arts training Jessie was:

Protected by the containing framework of the martial arts community, the structured relationship with the instructor, and the open communication between the instructor and therapist. Thus, it gave him enough space to think about his own traumatized thoughts and feelings and to verbalize them with the assistance of a therapist. After the community service Jessie did much better at school. He graduated from high school, went to college, and obtained an undergraduate degree before entering the military. He entered the military in a dangerous Special Forces division, which enabled him to continue to reflect in a leadership role there. It was a remarkable shift towards mentalizing in this otherwise sadistic, young person. (Twemlow et al., 2008, p. 17)

In this case example we can see how martial arts training in combination with psychotherapy was able to place this troubled youth on a productive life course and helped him to regulate his emotions and behaviors. The martial arts training also served him well in regard to being able to empathize with others and to become more sensitive and attuned to both himself and to those around him.

Case Vignette: "Paul"

Paul was a 30 year-old mentally retarded man with an IQ of 58. He would frequently have outbursts and act aggressively toward his mother. His mother and father had divorced and the mother was the sole custodian. The mother had a restraining order against the father so they were not able to directly communicate and work together on Paul's issues. Paul was exposed to domestic violence between his parents and this also had an impact on his overall development. Both parents expressed an interest in seeing their son live a normal and healthy life (Twemlow et al., 2008). Paul had long been bullied by his peers because of his disability. This severe bullying appears to have had a major impact on his personality and feelings toward others. At the time he began martial arts training he was already in therapy and was placed on psychiatric medication. He was mildly overweight and was known to be echolalic in his interactions with others. His body was tense and rigid and he did not like anyone touching him (Twemlow et al., 2008).

Paul loved to wear the karate uniform and it seems to give him a sense of comfort. During class it was common for Paul to act inappropriately, to speak out of turn, to run around, and to be disruptive. The instructor and the class would share the task of working with Paul and getting him to correct his behavior. The altruistic behaviors displayed by the martial arts class served to foster a mentalizing, other-directed mindset in both Paul and the other students in the class (Twemlow et al., 2008). Over time Paul learned to work within the rules of the dojo, to show respect to the other students and instructor, and how to adhere to group structure. Paul eventually developed a much improved self-defense skill set and appeared to always be very tired and relaxed after class. Eventually Paul was regarded as an integrated member of the larger group and was no longer viewed as a stigmatized mascot. His presence eventually had a major impact on the larger social climate in the dojo (Twemlow et al., 2008).

The martial arts training helped to instill a sense of community and altruism in both Paul and the other karate students. The dojo itself represents a community in miniature. By promoting an attitude of selflessness the martial arts training helped to connect Paul to others and to not view everyone else as cruel bullies. By bringing in martial arts training to complement his talk therapy and drug therapy Paul's life was significantly improved. Twemlow et al. (2008) highlight the resulting benefits of this therapeutic integration:

The combination of Depakote, Zyprexa and the martial arts training resulted in a 12-month period without physical outbursts at home. The parents proudly reported to the instructors their son's success at not exploding and following through on his responsibilities. The move here was that he was accepted, held contained and secure

in the attachment system that responded best to his needs to feel accepted and useful.

(p. 25)

This case example demonstrates the importance of having a securely attached containing

environment where the individual can feel safe and valued. In this example we can see how

important the social climate is and how integrating martial arts training with traditional

therapeutic methods dramatically improved the life and functioning of a seriously bullied

individual that struggled with both mental retardation and severe psychiatric problems

(Twemlow et al., 2008).

Brief Case Vignettes: Judo Training

Martial arts training can be incorporated into traditional therapy to modify social

behavior by utilizing techniques that focus on the body. Martial arts training is effective at

helping to treat mentally disturbed populations as can be seen in the following two brief

vignettes that describe the case of a young woman and a young man that both suffered from

psychotic disorders. Both of the following cases included traditional therapeutic approaches

that were in effect prior to the introduction of martial arts training. Judo was the form of martial

arts training that both patients received. This form of martial arts was incorporated into the

therapy because it was believed that judo principles have direct applications to psychotherapy

and are also relevant to teaching conflict-solving strategies (Gleser & Brown, 1988). Names

were not provided for either case so they will be referred to primarily by their respective gender

designations.

A 23 year-old man was referred to therapy during his third psychotic episode in four years. The psychotic episodes tended to be relatively short and responded to neuroleptic medication. The medication caused him many harmful side-effects and he was physically too weak to handle any level of tranquilizers. He was prescribed an eclectic treatment plan that included psychotherapy, physical therapy, and medication. It was deemed that physical activity would be useful to his treatment so he was placed on a plan that included running and weight training twice a week, judo twice a week, and psychotherapy twice a week. The young man was a yeshiva student so permission from the yeshiva director was obtained. Behavioral therapy was included in the treatment to help rid him of his fears and phobias (e.g., his fear of heights). Cognitive therapy was included in the treatment to help him with his depression. Analytic-oriented therapy was used to assist him in dealing with his unresolved oedipal conflict (Gleser & Brown, 1988).

Through the judo training and weight training he was able build his body and mentally ground himself. After six months of training he passed his first degree examinations in judo and at this time he presented with a much stronger sense of self-esteem and had a much better relationship with his family. He was also able to finally stand up to his authoritative father that wanted him to leave the yeshiva school and become a banker. At the three year follow up he continued in his judo training and had no further psychotic episodes and was no longer taking any medication. After being out of therapy for one year he was still training in judo and was free of any psychotic episodes. He reported that he was in good mental health and that his relationships were much better than before he began the training. At final follow up he had earned three judo degrees and was still in good health (Gleser & Brown, 1988).

This final brief case example demonstrates how incorporating judo training into traditional psychotherapeutic treatments can facilitate mental healing and behavioral improvement. A young woman who was diagnosed with paranoid schizophrenia was referred to a hospital after presenting with bizarre thinking and behavior. When she was going through a psychotic episode she would engage in inappropriate and dangerous behaviors. As an example, she would frequently enter men's restrooms and grab and pull on the men's sexual organs. These actions landed her in trouble on several occasions and she was even physical beaten on more than one occasion by the men whom she attacked (Gleser & Brown, 1988). For these reasons she was admitted into a hospital setting for further evaluation and psychotherapeutic treatment.

When she was not psychotic she was regarded by others as being quite shy. It was deemed by the clinical staff that physical exercise would be beneficial for her so running inside the hospital was incorporated into her treatment plan. After some time is was apparent that the running classes did not help to change or improve her psychotic behavior. It was then decided that martial arts training might be a good way to help her both mentally and physically. In these classes she was the only female. Gleser and Brown (1988) pointed out that, "In this case, judo, unlike the other treatments, 'yielded' to the girl's need for aggressive contact with men, while providing a socially acceptable framework in which she could achieve it. As a result she stopped attacking them" (p. 439). It appears that the judo training provided a therapeutic mode through the "yielding principle" and provided her a safe channel to rid herself of her aggressive urges and to stop attacking males. Both of these final vignettes focusing on two psychotic, regressed (and even one violent individual) demonstrate how core facets of judo and its integration into traditional mental health treatment can significantly advance psychotherapeutic outcomes (Gleser & Brown, 1988).

Chapter Eleven

Mind-Body Therapies as Natural and Healthy Alternatives

Alternatives to Drug Therapy

One of the motivating factors behind this analytic work is to demonstrate that natural and holistic treatment can, in many cases, be just as effective as potentially harmful and addictive drug therapies. We are seeing an enormous spike in the number of people taking prescription medication to try and treat their psychological and behavioral problems and this can be attributed largely to physicians trying to placate patients who were often not physically ill, but were instead suffering from problems with living (Pihl, 2009). Pihl carries this further by pointing out that all of this has been encouraged by the pharmaceutical industry that has vastly expanded its marketing campaigns and has broadened the definition of problems that require drug treatment. He viewed this as a "conspiracy" whereby the drug manufacturers made a major push to increase sales and profits through advertising and mass marketing that has had the main effect of turning attention away from the source of the problem and promote the belief that drugs solve problems and will improve one's overall quality of life and living (Pihl, 2009). This is misguided thinking. These prescribed drugs have their place but most psychological and behavioral problems require the treatment of the mind and the way that one is living.

This is not to say that there are no cases or disorders that could not be better served by including drug therapy into the treatment. The evidence is now clear that there are many circumstances where the problem is rooted in some type of biological malfunction. Preston, O'Neal, and Talaga (2013) pointed out that in certain broad and extremely heterogeneous clusters of disorders some may be primarily, or entirely, caused by biological factors (e.g., mood disorders). The authors also highlighted that convincing evidence now exists that some mental health illnesses are accompanied by, or caused by, neurochemical abnormalities. They argued that the failure to appropriately diagnose and treat these particular conditions will result in ineffective or only partially effective treatments. The result of this will be prolonged suffering for the patients and unnecessary increases in costs and financial burdens (Preston et al., 2013).

This work has not set out to deny that drug therapies have their place or that it does not serve to improve lives in many different situations. It is merely putting forth that in many cases there are safer and natural alternatives that are effective and that address directly the problems that so often cause psychological suffering and problems with living. In line with this perspective DiClemente (2003) informs us that prescription drugs often have a variable impact on the process of change and can undermine self-efficacy. He goes on to say that prescription medication may only enable patients to begin to make a change without having any type of decisional balance to support this change – or it may discourage the patient altogether from engaging in the necessary behavioral processes and components of change. This is not to argue that drug therapy cannot be useful across a range of mental health issues, it can be. This is to say that drug treatment is not merely a "magic bullet" and that those seeking permanent change and improvement in their lives need to address the underlying psychological and behavioral issues that afflict them. Even though Preston et al. (2013) writes on psychopharmacology and its effects in their work they also bring forth common concerns about overmedication and related problems associated with drug therapy, including nonadherence, suicidality, heavy financial burdens, and a wide range of dangerous side effects.

For the most part drugs do not cure clinical disorders or behavioral problems. Natural and healthy alternatives offer clients a relatively safe and long-lasting approach to mental health and well-being. As an example, a recent study released by Johns Hopkins University that conducted a meta-analysis of 47 clinical trials with over 3,500 participants concluded that mindfulness practices are just as effective at treating depression and anxiety as antidepressants (Goyal, et al., 2014). There were no harmful side effects arising from these trials and the effects appear to have generally improved the lives of the study participants. Zhaoming (2011) also put forth a related argument by presenting the evidence and findings that support mind-body therapies due to its minimal (or nonexistent) side effects, low costs, convenience, and therapeutic effectiveness. Zhaoming attributes the growing popularity of the various mind-body treatments to these various factors as well as the growing acceptance of them in clinical practice.

One of my basic motivations for this in-depth explication resides in the fact that we live in a culture that promotes drug use. The pharmaceutical companies have one basic agenda and that is to maximize profits. They spend an enormous amount of money each year lobbying, advertising, and marketing their products and they have done a fabulous job priming and conditioning the general population into believing that pills can fix or cure just about anything. As put forth above, medication is useful and necessary under certain conditions and for certain mental health illnesses. Based on what I have read in the literature and from what practicing clinicians have explained to me it seems clear that certain brain-based disorders (such as schizophrenia) are best treated through the proper use of medications. My argument was never to try and negate the importance of drug therapy. My position is that we have a serious epidemic in this country in regard to the use and abuse of prescription drugs (as well as illicit drugs). From everything that I have learned across my studies it seems fairly clear at this point in time that the rates of addiction and death across the nation that are related to prescription drugs is something that demands our serious attention.

It is my contention that the CAM therapies covered in this work are, in many cases, better alternatives than medication. When used in combination with CBT, mindfulness-based stress reduction, and/or mindfulness-based cognitive therapy, CAM modalities have been shown to be quite effective at treating a wide range of mental health issues and behavioral problems. As we have seen across this dissertation, integrating these natural and healthy approaches into psychological treatment has been shown to be quite useful in treating ailments such as depression, anxiety, addiction, eating disorders, healing from trauma, insomnia, stress-related disorders, externalizing behaviors, ADHD, and so forth. Although there may be some physical risk to certain CAM therapies (e.g., martial arts) there are no known side effects and they are not harmfully addictive. If anything, I would like to put forth that practicing CAM therapies such as yoga, martial arts, and meditation can actually result in positive and healthy lifelong "addictions" that carry across all aspects of life (e.g., healthy eating, healthy living, healthy relationships, longevity, etc.). We really have nothing to lose by giving these adjunctive modalities a chance, but we certainly have everything to gain from them.

Chapter Twelve

Concluding Thoughts on the Use and Value of Mind-Body Therapies

The primary purpose of this analytic exposition was to bring forth an abundance of relevant empirical data and findings to help demonstrate the usefulness of incorporating natural and healthy complementary mind-body therapies into clinical practice. As has been demonstrated throughout this work, it is not always sufficient to adhere exclusively to traditional talk therapies. The findings presented across this work demonstrated that it is often the case that incorporating mind-body practices with talk therapies, such as cognitive-behavioral therapy, can augment clinical outcomes. We have seen throughout this analytic investigation how a wide range of mental and behavioral health issues can be alleviated and improved through the use of certain complementary therapies, such as yoga, martial arts, biofeedback, meditation, progressive muscle relaxation, hypnosis, and so forth. We have also seen that combining certain complementary therapies with cognitive-behavioral therapy and certain mindfulness-based interventions (i.e., mindfulness-based stress reduction and mindfulness-based cognitive therapy) in psychological practice can prove to be very useful in serving clients. The basic goal of this integrative work was to push this narrative further and to promote a healthy and natural approach to psychological treatment.

This project has presented findings from case studies, randomized controlled research, nonrandomized controlled research, meta-analytic investigations, and so forth. It seems clear at this point that these adjunctive modalities can assist the larger psychotherapeutic community and help to better improve the lives of those that its serves. More research is needed to better demonstrate many of the causal mechanisms at play with these modalities but this will come in time. With this said, as of this writing the research evidence from both experimental designs and anecdotal observations seem to strongly support the claim that the highlighted mind-body therapies across this comprehensive monograph have broad clinical utility across a range of mental health and behavioral issues. Therapeutic integration with the spotlighted traditional therapies, CBT and the two MBI's, has also demonstrated clinical enhancement across many populations and diagnoses.

It is my recommendation that all psychotherapeutic practitioners learn about one or more of these complementary approaches. Learning about these various CAM therapies will help practitioners to be more effective in the overall healing process and in improving the lives and functioning of their clients. It is also my recommendation that clinical training programs offer more courses in CAM therapies so that future generations of clinicians will be educated and primed to think in terms of healing through natural and healthy remedies and treatments. These various mind-body therapies are widely available and are relatively inexpensive to learn and practice. For example, there are many youth centers, park and recreation centers, colleges and universities, and community centers that offer free yoga and martial arts classes. Many of these CAM therapies can be practiced in parks and in the privacy of one's home if necessary. The point here is that CAM therapies are generally accessible and affordable. I highly recommend that the psychotherapeutic community take advantage of these complementary therapeutic benefactions.

References

Abrams, B. (2001). Music, cancer, and immunity. *Journal of Oncology Nursing*, 5(5),

222-231.

Adkins, A. D., Singh, A. N., Winton, A. S., McKeegan, G. F., & Singh, J. (2010). Using a

mindfulness-based procedure in the community: Translating research to practice.

Journal of Child and Family Studies, 19, 175-183. doi: 10.1007/s10826-009-9348-9

Allison, D. B., & Faith, M. S. (1996). Hypnosis as an adjunct to cognitive-behavioral

psychotherapy for obesity: A meta-analytic reappraisal. *Journal of Consulting and

Clinical Psychology*. 64, 513-518. doi: 10.1037/0022-006X.64.3.513

American Psychological Association. (2007). *APA dictionary of psychology*. Washington,

DC: Author.

Arent, S. M., Landers, D. M., & Etnier, J. L. (2000). The effects of exercise on mood in older

adults: A meta-analytic review. *Journal of Aging and Physical Activity*, 8(4), 407-430.

Arias, A., Steinberg, K., Banga, A., & Trestman, R. (2006). Systematic review of the efficacy

of meditation techniques as treatments for medical illness. *Journal of Alternative and

Complementary Medicine*, 12, 817-832.

Arns, M., de Riddler, S., Strehl, U., Breteler, M., & Coenen, A. (2009). Efficacy of

neurofeedback treatment in ADHD: The effects of inattention, impulsivity, and

hyperactivity: A meta-analysis. *Clinical EEG and Neuroscience*, 40, 180-189. doi:

10.1177/155005940904000311

Astin, J. A. (1998). Why patients use alternative medicine. *Journal of the American Medical

Association*, 279, 1548-1553. doi: 10.1001/jama.279.19.1548

Baer, R. A. (2003). Mindfulness training as a clinical intervention: A conceptual and empirical review. *Clinical Psychology: Science and Practice*, 10, 125-143. doi: 10.1093/clipsy.bpg015

Baer, R. A., Fischer, S., & Huss, D. B. (2005). Mindfulness-based cognitive therapy applied to binge eating: A case study. *Cognitive and Behavioral Practice*, 123, 351-358.

Barbour, K. A., Edenfield, T. M., & Blumenthal, J. A. Exercise as a treatment for depression and other psychiatric disorders. *Journal of Cardiopuliminary Rehabilitation and Prevention*, 27(6), 359-367.

Barnes, J., Dong, C. Y., McRobbie, H., Walker, N., Mehta, M., & Stead, L. F. (2010). Hypnotherapy for smoking cessation. *Cochrane Database of Systematic Reviews*, 10, CD001008. Retrieved from http://onlinelibrary.wiley.com/doi/10.1002/14651858.CD001008.pub2/abstract

Barnett, J. E., & Shale, A. J. (2012). The integration of complementary and alternative medicine (CAM) into the practice of psychology: A vision for the future. *Professional Psychology: Research and Practice*, 43, 576-585. doi: 10.1037/a0028919

Barnett, J. E., Shale, A. J., Elkins, G., & Fisher, W. (2014). *Complementary and alternative medicine for psychologists: An essential resource*. Washington, DC: American Psychological Association.

Basler, A. J. (2011). Pilot study investigating the effects of Ayurvedic Abhyanga massage on subject stress experience. *The Journal of Alternative and Complementary Medicine*, 17, 435-440. doi: 10.1089/acm.2010.0281

Beck, J. S. (1995). *Cognitive therapy: Basics and beyond*. New York, NY: Guilford Press.

Bhargava, V., Hong, G. S., & Montalto, C. P. (2012). Use of practitioner-based and self-care complementary and alternative medicine in the United States: A demand for health perspective. *Family and Consumer Sciences Research Journal*, 41(1), 18-35. doi: 10.1111/j.1552-3934.2012.02126.x

Bhatia, T., Agarwal, A., Shah, G., Wood, J., Richard, J., Gur, R. E., Gur, R. C., Nimgaonkar, V. L., Mazumdar, S., & Deshpande, S. N. (2011). Adjunctive cognitive remediation for schizophrenia using yoga: an open, non-randomised trial. *International Journal of Neuropsychiatry*, 24, 91-100. doi: 10.1111/j.1601-5215.2011.00587.x

Bishop, S. R. (2002). What do we really know about mindfulness-based stress reduction? *Psychosomatic Medicine*, 64, 71-78.

Bogels, S., Hoogstad, B., van Dun, L., de Schutter, S., & Restifo, K. (2008). Mindfulness training for adolescents with externalizing disorders and their parents. *Behavioural and Cognitive Psychotherapy*, 36(2), 193-209. doi: 10.1017/S1352465808004190

Bootzin, R. R., & Stevens, S. J. (2005). Adolescents, substance abuse, and the treatment of insomnia and daytime sleepiness. *Clinical Psychology Review*, 25, 629-644.

Brotto, L. A., Mehak, L., & Kit, C. (2009). Yoga and sexual functioning: A review. *Journal of Sex and Marital Therapy*, 35, 378-390. doi: 10.1080/00926230903065955

Brown, K. W., Ryan, R. M., & Creswell, J. D. (2007). Mindfulness: Theoretical foundations and evidence for salutary effects. *Psychological Inquiry*, 18, 211-237.

Burschka, J. M., Kuene, P. M., Hofstadt-van Oy, U., Oschmann, P., & Kuhn, P. (2014). Mindfulness-based interventions in multiple sclerosis: Beneficial effects of Tai Chi on balance, coordination, fatigue, and depression. *BMC Neurology*, 14(1), 2-19. doi: 10.1186/S12883-014-0165-4

Caldwell, K., Harrison, M., Adams, M., Quinn, M. A., & Greeson, J. (2010). Developing mindfulness in college students through movement-based course: Effects on self-regulatory self-efficacy, mood, stress, and sleep quality. *Journal of American College Health*, 58(5), 433-442. doi: 10.1080/0744848-0903540481

Carei, T. R., Fyfe-Johnson, A. L., Breuner, C. C., & Brown, M. A. (2010). Randomized controlled clinical trial of yoga in the treatment of eating disorders. *Journal of Adolescent Health*, 46, 346-351. doi: 10.1016/j.jadohealth.2009.08.007

Carlson, L. E., Speca, M., Patel, K. D., & Goodey, E. (2003). Mindfulness-based stress reduction in relation to quality of life, mood, symptoms of stress, and immune parameters in breast and prostate cancer outpatients. *Psychosomatic Medicine*, 65(4), 571-581. doi: 10.1097/01.PSY.0000074003.35911.41

Carlson, L. E., Speca, M., Patel, K. D., & Goodey, E. (2004). Mindfulness-based stress reduction in relation to quality of life, mood, symptoms of stress and levels of cortisol, dehydroepiandrosterone sulfate (DHEAS) and melatonin in breast and prostate cancer outpatients. *Psychoneuroendocrinology*, 29(4), 448-474. doi: 10.1016/s0306-4350(03)00054-4

Carr, C., D'Ardenne, P., Sloboda, A., Scott, C. ,Wang, D., & Priebe, S. (2012). Group music therapy for patients with persistent post-traumatic stress disorder: An exploratory randomized controlled trial with mixed methods evaluation. *Psychology and Psychotherapy*, 85, 179-202. doi: 10.111/j.2044-8341.2011.02026.x

Carson, J. W., Carson, K. m., Porter, L. S., Keefe, F. J., Shaw, H., & Miller, J. M. (2007). Yoga for women with metastatic breast cancer: Results from a pilot study. *Journal of Pain and Symptom Management, 33*(3), 331-341. doi: 10.1016/j.painsymman.2006. 08.009

Casement, P. J. (1991). *Learning from the patient.* New York, NY: The Guilford Press.

Chambless, D. L., & Ollendick, T. (2002). Empirically supported psychological interventions: Controversies and evidence. *Annual Review of Psychology, 52*(1), 685-716.

Chen, Z. (2011). Integrative health. *Annals of the American Psychotherapy Association, 14*(1), 45-48.

Chesney, M., Ge, A., Gerber, L., Johnson, L. L., Mansky, P., & Ryan, M. (2006). Tai Chi Chuan: mind-body practice or exercise intervention? Studying the benefit for cancer survivors. *Integrative Cancer Therapies, 5*(3), 192-201.

Chiesa, A., Serretti, A. (2009). Mindfulness-based stress reduction for stress management in healthy people: A review and meta-analysis. *The Journal of Alternative and Complementary Therapy, 15,* 593-600. doi: 10.1089/acm.2008.0495

Chiesa, A, & Serretti, A. (2011). Mindfulness-based intervention for chronic pain: A systematic review of the evidence. *Journal of Alternative and Complementary Medicine, 17*(1), 83-93.

Coelho, H. F., Canter, P. H., & Ernst, E. (2013). Mindfulness-Based cognitive therapy: Evaluating current evidence and informing future research. *Psychology of Consciousness:Theory, Research, and Practice. 1*(S), 97-107. doi: 10.1037/2326-5523.I.S.97

Cramer, H., Lauche, R., Klose, P., Langhorst, J., & Dobos, G. (2013). Yoga for schizophrenia: A systematic review and meta-analysis. *BMC Psychiatry*, 13, 32-40. doi: 10.1186/1471-244X-13-32

Crane, R. (2009). *Mindfulness-based cognitive therapy: distinctive features*. New York, NY: Routledge.

Creamerm, P., Singh, B. B., Hochberg, M. C., & Berman, B. M. (2000). Sustained improvement produced by nonpharmacologic interventions in fibromyalgia. *Arthritis Care and Research*, 13(4), 198-204.

Cuddy, L. L., & Duffin, J. (2005). Music, memory, and Alzheimer's disease: Is music recognition spared in dementia, and how can it be assessed? *Medical Hypotheses*, 64, 229-240.

Culos-Reed, S. N., Carlson, L., Daroux, L. M., & Hately-Aldous, S. (2006). A pilot study of yoga for breast cancer survivors: Physical and psychological benefits. *Psycho-Oncology*, 15(10), 891-897.

Daley, A.J., Stokes-Lampard, H. J., & Macarthur, C. (2009). Exercise to reduce vasomotor and other menopausal symptoms: A review. *Maturitas*, 63(3), 176-180. doi: 10.1016/j.maturitas.2009.02.004

D'Angelo, A. (2013). Building bridges. The appropriation of yoga therapy to the emotional dimensions of ailments. *Brazilian Journal of Sociology of Emotion*, 12(3), 321-360.

Davidson, R., Kabat-Zinn, J., Schumacher, J., Rosenkranz, M. Muller, D., & Santorelli, D. (2003). Alterations in brain and immune function produced by mindfulness meditation. *Psychosomatic Meditation*, 65, 564-570.

DeAngelis, T. (2013). A natural fit. *Monitor on Psychology*, 44(8), 56-59.

Dhikav, V., Karmarkar, G., Gupta, R., Verma, M., Gupta, R., Gupta, S., & Anand, K. S. (2010). Yoga in female sexual functions. *Journal of Sexual Medicine*, 7(2), 964-970. doi: 10.1111/j.1743-6109.2009.01580.x

DiClemente, C. C. (2003). *Addiction and change: How addictions develop and addicted people recover*. New York, NY: The Guilford Press.

Dimidjian, S., & Linehan, M. M. (2003). Defining an agenda for future research on the clinical application of mindfulness practice. *Clinical Psychology: Science and Practice*, 10, 166-171. doi: 10.1093/clipsy.bpg019

DiStasio, S. A. (2007). Integrating yoga into cancer care. *Clinical Journal of Oncology Nursing*. 12(1), 125-131. doi: 10.1188/08.CJON.125-130

Ditto, B., Eclache, M., & Goldman, N. (2006). Short-term autonomic and cardiovascular effects of mindfulness body scan meditation. *Annals of Behavioural Medicine*, 32, 227-234.

Edenfield, T. M., & Saeed, S. A. (2012). An update on mindfulness meditation as a self-help treatment for anxiety and depression. *Psychology Research and Behavior Management*, 5,131-141. doi: 10.2147/PRBM.S34937

Elkins, G., Marcus, J., Bates, J., Rajab, H. M., & Cook, T. (2006). Intensive hypnotherapy for smoking cessation: A prospective study. *International Journal of Clinical and Experimental Hypnosis*, 54, 303-315. doi: 10.1080/00207140600689512

Engelman, S. R. (2013). Palliative care and use of animal-assisted therapy. *Omega*, 67(1-2), 63-67. doi: http://dx.doi.org/a0.2190/OM.67.1-2.g

Evans, S., Ferrando, S., Findler, M., Stowell, C., Smart, C., & Haglin, D. (2008). Mindfulness-based cognitive therapy for generalized anxiety disorder. *Journal of Anxiety Disorders*, 22, 716-721.

Field, T. (2009). *Complementary and alternative therapies research*. Washington, DC: American Psychological Association.

Field, T. (2011). Yoga clinical research review. *Complementary therapies in clinical practice*,17(1), 1-8. doi: 10.1016/j.ctcp.2010.09.007

Field, T. (2012). Exercise research on children and adolescents. *Complementary Therapies in Clinical Practice*, 18(1), 54-69. doi: 10.1016/j.ctcp.2011.04.002

Forbes, D., Phelps, A., & McHugh, T. (2001). Treatment of combat-related nightmares using imagery rehearsal: A pilot study. *Journal of Traumatic Stress*, 14, 433-442.

Fouladbakhsh, J. M., & Stommel, M. (2010). Gender, symptom experience, and use of complementary and alternative medicine practices among cancer survivors in the U.S. cancer population. *Oncology Nursing Forum*, 37(1), 7-15. doi: 10.1188/10.ONF.E7-E15

Freeman, L. (2009). *Mosby's complementary & alternative medicine: A research-based approach* (3rd ed.). St. Louis, MO: Mosby.

Fuller, J. R. (1998). Martial arts and psychological health. *British Journal of Medical Psychology*, 61, 317-328.

Galantino, M. L., Shepard, K., Krafft, L., LaPerriere, A., Ducette, J., Sorbelo, A., … Barnish, M. (2005). The effect of group aerobic exercise and Tai Chi on functional outcomes and quality of life for persons living with acquired immunodeficiency syndrome. *The Journal of Alternative and Complementary Medicine*, 11(6), 1085-1092.

Garland, E.L. (2007). The meaning of mindfulness: A second-order cybernetics of stress, metacognition, and coping. *Complementary Health Practice Review*, 12, 15-30.

Gemmell, C., & Leathem, J. M. (2006). A study investigating the effects of Tai Chi Chuan: Individuals with traumatic brain injury compared to controls. *Brain Injury*, 20(2), 151-156. doi: 10.1080/0269905050442998

Ghoncheh, S., & Smith, J. C. (2004). Progressive muscle relaxation, yoga, stretching, and ABC relaxation theory. *Journal of Clinical Psychology*, 60(1), 131-136. doi: 10.1002/jclp.10194

Gleser, J., Brown, & Brown, P. (1988). Judo principles and practices: Applications to conflict-solving strategies in psychotherapy. *American Journal of Psychotherapy*, 42(3), 437-447.

Goldin, P. R., & Gross, J. J. (2010). Effects of mindfulness-based stress reduction (MBSR) on emotion regulation in social anxiety disorder. *Emotion*, 10, 83-91. doi: 10.1037/a0018441

Goyal, M., Singh, S., Sibinga, E. M., Gould, M. F., Rowland-Seymour, A., Sharma, R.,… Haythornthwaite, J. A. (2014). Meditation programs for psychological stress and well-being: A systematic review and meta-analysis. *Journal of the American Medical Association Internal Medicine*, 174(3), 357-368. doi: 10.1001/ jamainternmed. 2013.13018

Granath, J., Ingvarsson, S., von Thiele, U., Lundberg, U. (2006). Stress management: A randomized study of cognitive behavioural therapy and yoga. *Cognitive Behaviour Therapy*, 35(1), 3-10.

Green, J. P., & Lynn, S. J. (2000). Hypnosis and suggestion-based approaches to smoking cessation: An examination of the evidence. *International Journal of Clinical and Experimental Hypnosis*, 48, 191-198.

Grossman, P., Kappos, L., Gensicke, H., D'Souza, M., Mohr, D. C., Penner, I. K., & Steiner, C. (2010). MS quality of life, depression, and fatigue improve after mindfulness training: A randomized trial. *Neurology*, 75(13), 1141-1149. doi: 10.1212/ WNL.0b013e3181f4d8od

Guerevich, M. I., Duckworth, D., Imhof, J. E., & Katz, J. L. (1996). Is auricular acupuncture beneficial in the inpatient treatment of substance-abuse patients? A pilot study. *Journal of Substance Abuse Treatment*, 13, 165-171.

Guthrie, S. R. (1995). Liberating the Amazon: Feminism in the martial arts. *Women & Therapy*, 16(2-3), 107-119. doi: 10.1300/J01Sv16n02-12

Hammond, D. C. (2005). Neurofeedback with anxiety and affective disorders. *Child and Adolescent Psychiatric Clinics of North America*, 14, 105-123.

Hammond, D. C. (2007). Review of the efficacy of clinical hypnosis with headaches and migraines. *International Journal of Clinical and Experimental Hypnosis*, 55, 207-219.

Harris, A. H., Cronkite, R., & Moos, R. (2006). Physical activity, exercise coping, and depression in a 10-year cohort study of depressed patients. *Journal of Affective Disorders*, 93(3), 79-85. doi: 10.1016/j.jad.2006.02.013

Haydicky, J. (2010). *Mindfulness training for adolescents with learning abilities*. Retrieved from Creative Commons. (UMI no. 1807/24228)

Haydicky, J., Wiener, J., Badali, P., Milligan, K., & Ducharme, J. M. (2012). Evaluation of a mindfulness-based intervention for adolescents with learning disabilities and co-occurring ADHD and anxiety. *Mindfulness*, 3(2), 151-164. doi: 10/1007/s12671-012-0089-2

Hernandez-Ruiz, E. (2005). Effect of music therapy on the anxiety levels and sleep patterns of abused women in shelters. *Journal of Music Therapy*, 42, 140-158.

Hoffman, S. G., & Smits, J. A. (2008). Cognitive-behavioral therapy for adult anxiety disorders: A meta-analysis of randomized placebo-controlled trials. *Journal of Clinical Psychiatry*, 69(4), 612-632. doi: 10.4088/JCP.v69n0415

Hoffman, S. G., Sawyer, A. T., Witt, A. A., & Oh, D. (2010). The effect of mindfulness–based therapy on anxiety and depression: A meat-analytic review. *Journal of Consulting and Clinical Psychology*, 78(2), 169-183. doi: 10.1037/a0018555

Holmes, A. E., & Matthews, A. (2005). Mental imagery and emotion: A special relationship? *Emotion*, 5, 489-497.

Holtmann, M., & Stadler, C. (2006). Electroencephalographic biofeedback for the treatment of attention-deficit hyperactivity disorder in childhood and adolescence. *Expert Review of Neurotherapies*, 6, 533-540.

Hsu, W. C., & Lai, H. L. (2004). Effects of music on major depression in psychiatric inpatients. *Archives of Psychiatric Nursing*, 18, 193-199.

Huang, T. T., Yang, L. H., & Liu, C. Y. (2011). Reducing the fear of falling among community-dwelling elderly adults through cognitive-behavioural strategies and intense Tai Chi exercise: A randomized controlled trial. *Journal of Advanced Nursing*, 67(5), 961-971. doi: 10.1111/J.1365-2648.2010.05553.x

Iaboni, A. & Flint, A. J. (2013). The complex interplay of depression and falls in older adults: A clinical review. *American Journal of Geriatric Psychiatry*, 21(5), 484-492. doi: 10.1016/j.jagp.2013.01.008

International Association of Yoga Therapists. (2014). *About the International Association of Yoga Therapists*. Retrieved from http://www.iayt.org/?page=LearnAbout

Jacobson, J. S., & Verret, W. J. (2001). Complementary and alternative therapy for breast cancer: The evidence so far. *Cancer Practice*, 9(6), 307-310.

Jamieson, G. A. (2012). Contemporary clinical hypnosis: Theory and practice. *Australian Journal of Clinical & Experimental Hypnosis*, 40(1), 13-19.

Jones, M. C., & Johnston, D. W. (2000). Reducing distress in first level and student nurses: A review of the applied stress management literature. *Journal of Advanced Nursing*, 32, 66-74.

Jordan, J. B. (2006). Acupuncture treatment for opiate addiction: A systematic review. *Journal of Substance Abuse Treatment*, 30, 309-314.doi: 10.1016/j.jsat.2006.02.005

Justina, L. Y., & Man, T. C. (2014). A randomized trial comparing Tai Chi with and without cognitive-behavioral intervention (CBI) to reduce fear of falling in community-dwelling elderly people. *Archives of Gerontology and Geriatrics*, 59, 317-325. doi: 10. 1016/j.archger.2014.05.008

Kabat-Zinn, J., Massion, A. O., Kristeller, J., & Peterson, L. G. (1992). Effectiveness of a meditation-based stress reduction program in the treatment of anxiety disorders. *American Journal of Psychiatry*, 149(7), 936-943.

Kabat-Zinn, J. (2003). Mindfulness-based interventions in context: Past, present, and future. *Clinical Psychology: Science and Practice*, 10, 144-158.

Kalb, L. M., & Loeber, R. (2003). Child disobedience and noncompliance: A review. *Pediatrics*, 111, 641-652.

Kaley-Isley, L. C., Peterson, J., Fischer, C., & Peterson, E. (2010). Yoga as a complementary therapy for children and adolescents: A guide for clinicians. *Psychiatry*, 7(8), 20-32.

Kaliappan, K. V. (1998). Personality development as deviance and social control. *International Sociological Association*, 7(2), 82-85.

Karavidas, M. K., Lehrer, P. M., Vaschillo, E., Vaschillo, B., Marin, H., & Buyske, S. (2007). Preliminary results of an open label study of heart rate variability biofeedback for the treatment of major depression. *Applied Psychophysiology and Biofeedback*, 32, 19-30.

Karst, M., Winterhalter, M., Munte, S., Francki, B., Hondronikos, A.,....Eckardt, A. (2007). Auricular acupuncture for dental anxiety: A randomized controlled trial. *Anesthesia and Analgesia*, 104, 295-300.

Kaushik, R., Kaushik, R. M., Mahajan, S. K., & Rajesh, V. (2005). Biofeedback assisted diaphragmatic breathing and systematic relaxation versus propranolol in long term prophylaxis of migraine. *Complementary Therapies in Medicine*, 13, 165-174.

Kerr, C. (2002). Translating "mind-in-body": Two models of patient experience underlying a randomized controlled trial of Qigong. *Culture, Medicine, & Psychiatry*, 26(4), 419-447.

Kessler, R. C., Soukup, J., & Davis, R. B. (2001). The use of complementary and alternative therapies to treat anxiety and depression in the United States. *American Journal of Psychiatry*, 158(2), 289-294.

Khalsa, S. B. (2004). Treatment of chronic insomnia with yoga: A preliminary study with sleep-wake diaries. *Applied Psychophysiology and Biofeedback*, 29, 269-278.

Khalsa, M. K., Greiner-Ferris, J. M., Hofmann, S. G., & Khalsa, S. B. (2014). Yoga-enhanced cognitive behavioural therapy (y-cbt) for anxiety management: A pilot study. *Clinical Psychology & Psychotherapy*, 7, 1-8. doi: 10.1002/cpp.1902

Kiecolt-Glaser, J. K., Christian, L., Preston, H., Houts, C. R., Malarkey, W. B., Emery, C. F., & Glaser, R. (2010). Stress, inflammation, and yoga practice. *Psychosomatic Medicine*, 72, 113-121. doi: 10.0033-3174/10/7202-0113

Kinser, P. A., Bourguignon, C., Taylor, A. G., & Steeves, R. (2013). "A feeling of connectedness": perspectives on a gentle yoga intervention for women with major depression. *Issues in Mental Health Nursing*, 34(6), 402-411. doi: 10.3109/01612840. 2012.762959

Kirkwood, G., Rampes, H., Tuffrey, V., Richardson, J., & Pilkington, K. (2005). Yoga for anxiety: A systematic review of the research evidence. *British Journal of Sports Medicine*, 39(12), 884-891. doi: 10.1136/bjsm.2005. 018069

Kirsch, I. (1996). Hypnotic enhancement of cognitive-behavioral weight loss treatments— another meta-reanalysis. *Journal of Consulting Clinical Psychology*, 64(3), 517-523. doi: 10.1037/0022-006X.64.3.517

Kirsch, I., Montgomery, G., & Sapirstein, G. (1996). Hypnosis as an adjunct to cognitive-behavioral psychotherapy: a meta-analysis. *Journal of Consulting and Clinical Psychology*, 63, 214-221.

Ko, G., Tsang, P., & Chang, H. (2006). A 10-week Tai-Chi program improved the blood pressure, lipid profile and SF-36 scores in Hong Kong Chinese women. *Medical Science Monitor*, 12, 196-199.

Kozasa, E. H., Hachul, H., Monson, C., Pinto, L., Garcia, M. C., de Araujo Moraes Mello, L. E., & Tufik, S. (2010). Mind-body interventions for the treatment of insomnia: A review. *Brazilian Journal of Psychiatry*, 32(4), 437-443.

Krisanaprakornkit, T., Krisanaprakornkit, W., Piyavhatkul, N., & Laopaiboon, M. (2006). Meditation therapy for anxiety disorders. *Cochrane Database of Systematic Reviews*, 25(1), CD004998. doi: 10.1002/14651858.CD004998.pub2

Krisanaprakornkit, T., Ngamjarus, C., Witoonchart, C., & Piyavhatkul, N. (2010). Meditation therapies for attention-deficit/hyperactivity disorder (ADHD). *Cochrane Database of Systematic Review*, 16(6), CD006507. doi: 10.1002/14651858.CD006507.pub2

Kuan-Yin, L., Yu-Ting, H., King-Jen, C., Heui-Fen, L., & Jau-Yih, T. (2011). Effects of yoga on psychological health, quality of life, and physical health of patients with cancer: A meta-analysis. *Evidence-Based Complementary and Alternative Medicine*, 11, 1-12. doi: 10.1155/2011/659876

Kubsch, S. M., Neveau, T., & Vandertie, K. (2000). Effects of cutaneous stimulation on pain reduction in emergency department patients. *Complementary Therapies in Nursing and Midwifery*, 11, 28-33.

Lane, J. D., Seskevich, J. E., & Peiper, C. F. (2007). Brief meditation training can improve perceived stress and negative mood. *Alternative Therapies in Health and Medicine*, 13, 38-44.

Lantz, J. (2002). Family development and the martial arts: A phenomenological study. *Contemporary Family Therapy*, 24(4), 565-580.

Lau, M. A., & McMain, S. F. (2005). Integrating mindfulness meditation with cognitive and behavioural therapies: The challenge of combining acceptance and change-based strategies. *Canadian Journal of Psychiatry*, 50(13), 863-869.

Lau, M. A., Segal, Z. V., & Williams, J. M. (2004). Teasdale's differential activation hypothesis: Implications for mechanisms of depressive relapse and suicidal behaviour. *Behaviour Research and Therapy*, 42, 1001-1017.

Lee, M. S., Pittler, M. H., & Ernst, E. (2007). Is Tai Chi an effective adjunct in cancer care? A systematic review of controlled clinical trials. *Supportive Care in Cancer*, 15(6), 597-601.

Lee, S. H., Ahn, S. C., Lee, Y. J., Choi, T. K., Yook, K. H., & Suh, S. Y. (2007). Effectiveness of a meditation based stress management program as an adjunct to pharmacotherapy in patients with anxiety disorder. *Journal of Psychosomatic Research*, 62, 189-195.

Lengacher, C. A. (2012). Mindfulness-based cognitive therapy for cancer. *Psycho-Oncology*, 21(8), 9-11. doi: 10.1002/pon.3129

Levine, A. S., & Balk, J. L. (2012). Yoga and quality-of-life improvement in patients with breast cancer: a literature review. *International Journal of Yoga Therapy*, 22, 95-99.

Lewinsohn, P. M., Allen, N. B., Seeley, J. R., Gotlib, I. H. (1999). First onset versus recurrence of depression: Differential processes of psychosocial risk. *Journal of Abnormal Psychology*, 108, 483-489.

Li, F., Fisher, K., Harmer, P., Irbe, D., Tearse, R., & Weimer, C. (2004). Tai Chi and self-rated quality of sleep and daytime sleepiness in older adults: A randomized controlled trial. *Journal of the American Geriatric Society*, 5, 892-900.

Libby, D. J., Reddy, F., Pilver, C. E., & Desai, R. A. (2012). The use of yoga in specialized VA PTSD treatment programs. *International Journal of Yoga Therapy*, 22, 79-87.

Lin, K. Y., Hu, Y. T., Chang, K. J., Lin, H. F., & Tsauo, J. Y. (2011). Effects of yoga on psychological health, quality of life, and physical health of patients with cancer: A meta-analysis. *Evidence-Based Complementary and Alternative Medicine*, ID 659876, 1-12. doi: 10.1155/2011/659876

Liossi, C., & Hatira, P. (2003). Clinical hypnosis in the alleviation of procedure-related pain in pediatric oncology patients. *International Journal of Clinical Experimental Hypnosis*, 51, 4-28.

Lipton, D. S., Brewington, V., & Smith, M. (1994). Acupuncture for crack-cocaine detoxification: Experimental evaluation of efficacy. *Journal of Substance Abuse Treatment*, 11(3), 205-213.

Lothes II, J., Hakan, R., & Kassab, K. (2013). Aikido experience and its relation to mindfulness: A two-part study. *Perceptual and Motor Skills*, 116 (1), 30-39. doi: 10. 2466/22.23.PMS.116.1.30-39

Lykins, E. L. B., & Baer, R. A. (2009). Psychological functioning in a sample of long-term practitioners of mindfulness meditation. *Journal of Cognitive Psychotherapy*, 23, 226-241. doi: 10.1891/0889-8391.23.3.226

Ma, H. S., & Teasdale, J. D. (2004). Mindfulness-based cognitive therapy for depression: Replication and exploration of differential relapse prevention effects. *Journal of Consulting and Clinical Psychology*, 72(1). 31-40. doi: 10.1037/0022-006X.72.1.31

Mace, C. (2008). *Mindfulness and mental health: Therapy, theory, and science*. New York, NY: Routledge.

Manicavasagar, V., Perich, T., & Parker, G. (2012). Mindfulness-based cognitive therapy vs cognitive behaviour therapy as a treatment for non-melancholic depression. *Journal of Affective Disorders*, 130, 138-144.

Mansky, P. J., & Wallerstedt, D. B. (2006). Complementary medicine in palliative care and cancer symptom management. *Cancer Journal*, 12(5), 425-431.

Martin, D. G. (2011). *Counseling and therapy skills* (3rd ed.). Long Grove, IL: Waveland Press.

Martinsen, E. W. (2008). Physical activity in the prevention and treatment of anxiety and depression. *Nordic Journal of Psychiatry*, 62, 25-29. doi: 10.1080/08039480802315640

McCallie, M., Blum, C., & Hood, J. (2006). *Progressive muscle relaxation. Journal of Human Behavior in the Social Environment*, 13, 51-66.

McKenna, M. (2001). The application of Tai Chi Chuan in rehabilitation and preventative care of the geriatric population. *Physical and Occupational Therapy in Geriatrics*, 18(4), 23-28.

Mehta, P. & Sharma, M. (2010). Yoga as a complementary therapy for clinical depression. *Complementary Health Practices Review*, 15(3), 156-170.doi:10.1177/ 15332101110387-405

Meister, I. G., Krings, T., Foltys, H., Boroojerdi, B., Muller, M., Topper, R., & Thron, A. (2004). Playing piano in the mind: An fMRI study on music imagery and performance in pianists. *Brain Research*, 19, 219-228.

Menzies, V., Taylor, A. L. (2004). The idea of imagination: An analysis of "imagery," *Advances*, 20(2), 4-13.

Michalsen, A., P., Grossman, P., Acil, A., Langhorst, J., Ludtke, R., & Esch, T. (2005). Rapid stress reduction and anxiolysis among distressed women as a consequence of a three-month intensive yoga program. *Medicine and Science Monitoring*, 11, 555-561.

Michalsen, A. P., Jeitler, M., Brunnhuber, S., Ludtke, R, Bussing, A., Musial., F....Kessler, C. (2012). Iyengar yoga for distressed women: A 3-armed randomized controlled trial. *Evidence-Based Complementary Alternative Medicine*, 408727, 1-9. doi: 10.1155/ 2012/408727

Mills, N. & Allen, J. (2000). Mindfulness of movement as a coping strategy in multiple sclerosis: A pilot study. *General Hospital Psychiatry*, 22(6), 425-431.

Moore, N. C. (2000). A review of EEG biofeedback treatment of anxiety disorders. *Clinical EEG*, 31, 1-6.

Monastra, V. J. (2005). Electroencephalographic biofeedback (neurotherapy) as a treatment for attention deficit hyperactivity disorder: Rationale and empirical foundation. *Child and Adolescent Psychiatric Clinics of North America*, 14, 55-82.

Moraska, A., Pollini, R. A., Boulanger, K., Brooks, M. Z., & Teitlebaum. L. (2010).

Physiological adjustments to stress measures following massage therapy. *Evidence-Based Complementary and Alternative Medicine*, 7, 409-418. doi: 10.1093/ecam/nen29

Morley, S., Eccleston, C., & Williams, A. (1999). Systematic review and meta-analysis of

randomized controlled trials of cognitive behaviour therapy and behaviour for chronic

pain in adults, excluding headache. *Pain*, 80, 1-13. doi: 10.1016/S0304-39590025-3

Moyer, C. A., Rounds, J., & Hannum, J. W. (2004). A meta-analysis of massage therapy

research. *Psychological Bulletin*, 130, 3-18. doi: 10.1037/0033-2909.130.1.3

Muir, J. M. (2012). Chiropractic management of a patient with symptoms of attention-deficit/

hyperactivity disorder. *Journal of Chiropractic Medicine*, 11, 221-224. doi: 10.1016/

j.jcm.2011.10.009

Murphy, L. R. (1996). Stress management in work settings: A Critical review of the health

effects. *American Journal of Health Promotion*, 11, 112-135.

Nahin, R. L., Barnes, P. M., Stussman, B. J., & Bloom, B. (2009). Costs of complementary

and alternative medicine (CAM) and frequency of visits to CAM practitioners: United

States, 2007. *National Health Statistics Report*, 30(18), 1-14.

National Institutes of Health, National Canter for Complementary and Alternative Medicine.

(2010). *Mind-body medicine practices in complementary and alternative medicine.*

Retrieved from http://report.nih.gov/nihfactsheets/Pdfs/MindBodyMedicinePracticesin

ComplementaryandAlternativeMedicine%28NCCAM29.pdf

Orme-Johnson, D. W., Schneider, R. H., Son, Y. D., Nidich, S., & Cho, Z. H. (2006).

Neuroimaging of meditation's effects on brain reactivity to pain. *Neuroreport*, 17,

1359-1363. doi: 10.1097/01.wnr.0000233094.67289.a8

Ott, M. J. (2002). Mindfulness meditation in pediatric clinical practice. *Pediatric Nursing,* 28, 487-491.

Otto, M., Norris, R., & Bauer-Wu, S. (2006). Mindfulness meditation for oncology patients: A discussion and critical review. *Integrated Cancer Therapy,* 5, 98-108.

Palermo, M. T., Di Luigi, M., Dal Forno, G., Dominici, C., Vicomandi, D., Sambucioni, A., Proietti, L., & Pasqualetti, P. (2006). Externalizing and oppositional behaviors and karate-do: The way of crime prevention: A pilot study. *International Journal of Offender Therapy and Comparative Criminology,* 50(6), 654-660.doi: 10.1177/03066-224X06293522

Park, C. (2013). Mind-Body CAM interventions: Current status and considerations for integration into clinical psychology. *Journal of Clinical Psychology,* 69(1), 45-63. doi: 10.1002/jclp.21910

Pawlow, L., & Jones, G. (2005). The impact of abbreviated progressive muscle relaxation on salivary cortisol and salivary immunoglobulin A (sIgA). *Applied Psychophysiology and Biofeedback,* 30, 375-387.

Penninx, B. W., Rejeski, J. W., Pandya, J., Miller, M. E., Di Bari, M., Applegate, W. B., & Pahor, M. (2002). Exercise and depressive symptoms: A comparison of aerobic and resistance exercise effects on emotional and physical function in older persons with high and low depressive symptomatology. *The Journals of Gerontology: Series B: Psychological Sciences and Social Sciences,* 5(2), 124-132. doi: 10.1093/geronb/57.2. P124

Phelps, A., & McHugh, T. (2001). Treatment of combat-related nightmares using imagery rehearsal: A pilot study. *Journal of Traumatic Stress,* 14(2), 433-441.

Pihl, R. O. (2009). Substance abuse: Etiological considerations. In P. Blaney & T. Millon (Eds.), *Oxford textbook of psychopathology* (2nd ed., pp. 253-279). New York, NY: Oxford University Press.

Pippa, L., Manzoli, L., Corti, I., Congedo, G., Romanazzi, L., & Parruti, G. (2007). Functional capacity after traditional Chinese medicine (Qigong) training in patients with chronic atrial fibrillation: A randomized controlled trial. *Preventative Cardiology*, 10, 22-25.

Plante, T. G. (2011). *Contemporary clinical psychology* (3rd ed.). Hoboken, NJ: Wiley.

Pratt, K. M. (2013). Evaluating complementary and alternative medicine (CAM) utilization in a college sample: a multisite application of the sociobehavioral model of healthcare utilization. *Journal of Alternative and Complementary Medicine*, 74(3), 192-199.

Preston, J. D., O'Neal, J. H., & Talaga, M. C. (2013). *Handbook of clinical psychopharmacology for therapists* (7th ed.). Oakland, CA: New Harbinger Publications, Inc.

Raschetti, R., Menniti-Ippolito, F., Forcella, E., Bianchi, C. (2005). Complementary and alternative medicine in the scientific literature. *Journal of Alternative and Complementary Medicine*, 11, 209-215.

Rausch, S., Gramling, S., & Auerbach, S. (2006). Effects of a single session of large-group meditation and progressive muscle relaxation training on stress reduction, reactivity, and recovery. *International Journal of Stress Management*, 13, 273-290.

Reynes, E., & Lorant, J. (2004). Competitive martial arts and aggressiveness: A 2-yr longitudinal study among young boys. *Perceptual and Motor Skills*, 98, 103-115.

Riley, D. (2003). Hatha yoga and the treatment of illness. *Alternative Therapies in Health and Medicine*, 10, 20-23.

Saadat, H., Drummond-Lewis. J., Maranets, I., Kaplan, D., Saadat, A., Wang, S., & Kain, Z. (2006). Hypnosis reduces preoperative anxiety in adult patients. *Anesthesia and Analgesia*, 102, 1394-1396.

Saeed, S. A., Antonacci, D. J., & Bloch, R. M. (2010). Exercise, yoga, and meditation for depressive and anxiety disorders. *American Family Physician*, 81(8), 981-986.

Salmon, P., Lush, E., Jablonski, M., Sephton, S. E. (2009). Yoga and mindfulness: Clinical aspects of an ancient mind/body practice. *Cognitive and Behavioral Practice*, 16, 59-72.

Schoenberger, N. E. (2000). Research on hypnosis as an adjunct to cognitive-behavioral psychotherapy. *International Journal of Clinical Experimental Hypnosis*, 48(2), 154-161.

Sharma, S., & Kaur, L. (2006). Hypnosis and pain management. *Nursing Journal of India*, 97, 127-131.

Shifflett, C. M. (1999). *Aikido exercises for teaching and training*. Berkeley, CA: North Atlantic Books.

Shigaki, C. L., Glass, B., & Schopp, L. H. (2006). Mindfulness-based stress reduction in medical settings. *Journal of Clinical Psychology in Medical Settings*, 13, 209-216.

Simkin, D. R., & Black, N. B. (2014). Meditation and mindfulness in clinical practice. *Child and Adolescent Psychiatric Clinics of North America*, 23(3), 487-534. doi: 10. 1016/j.chc.2014.03.002

Simpson, R., Booth, J., Lawrence, M., Byrne, S., Mair, F., & Mercer, S. (2014). Mindfulness based interventions in multiple sclerosis – a systematic review. *BMC Neurology*, 14, 1-19. doi: 10.1186/1471-2377-14-15

Singh, N. A., Clements, K. M., & Fiatarone-Singh, M. A. (2001). The efficacy of exercise as a long-term antidepressant in elderly subjects: A randomized, controlled trial. *Journal of Gerontology*, 56(8), 497-504.

Sloman, R. (2002). Relaxation and imagery for anxiety and depression control in community patients with advanced cancer. *Cancer Nursing* 25, 432-435.

Smith, J. E., Richardson, J., Hoffman, C., & Pilkington, K. (2005). Mindfulness-based stress reduction as supportive therapy in cancer care: Systematic review. *Journal of Advanced Nursing*, 52(3), 315-327.

Smith, K. B., & Pukall, C. F. (2009). An evidence-based review of yoga as a complementary intervention for patients with cancer. *Psychooncology*, 18(5), 465-475.

Spahn, G., Lehmann, N., Franken, U., Paul, A., Langhorst, J., Michalsen, A., & Dobos, G. J. (2003). Improvement of fatigue and role function of cancer patients after an outpatient integrative mind-body intervention. *Focus on Alternative and Complementary Therapies*, 8(4), 540-549.

Speca, M., Carlson, L., Goodey, E., & Angen, M. (2000). A randomized wait-list controlled clinical trial: The effect of a mindfulness-based stress reduction program on mood and symptoms of stress in cancer outpatients. *Psychosomatic Medicine*, 62, 613-622.

Spiegler, M. D., & Guvremont, D. C. (2010). *Contemporary behavior therapy*. Belmont, CA: Wadsworth.

Spinazzola, J., Rhodes, A. M., Emerson, D., Earle, E., & Monroe, K. (2011). Application of yoga in residential treatment of traumatized youth. *Journal of the American Psychiatric Nurses Association*, 17(6), 431-444. doi: 10.1177/1078390311418359

Sun, J., Buys, N., & Jayasinghe, R. (2014). Effects of community-based meditative Tai Chi programme on improving quality of life, physical and mental health in chronic heart-failure participants. *Aging and Mental Health*, 18(3), 289-295. doi: 10.1080/13607863. 2013.875120

Syrjala, K. L., Cummings, C., & Donaldson, G. W. (1992). Hypnosis or cognitive behavioral training for the reduction of pain and nausea during cancer treatment: A controlled clinical trial. *Pain*, 48, 137-146.

Tahiri, M., Motillo, S., Joseph, L., Pilote, L., & Eisenberg, M. J. (2012). Alternative smoking cessation aids: A meta-analysis of randomized controlled trials. The *American Journal of Medicine*, 125, 576-584. doi: 10.1016/j.amjmed.2011.09.028

Tavee, J. Rensel, M., Planchon, S., Stone, L. (2011). Effects of meditation on pain and quality of life in multiple sclerosis and peripheral neuropathy: A controlled study. *International Journal of Multiple Sclerosis Care*, 13(2), 163-168.

Taylor-Pilliae, R. E., Haskell, W. L., Waters, C. M., & Froelicher, E. S. (2006). Change in perceived psychosocial status following a 12-week Tai Chi exercise programme. *Journal of Advanced Nursing*, 54(3), 313-329. doi: 10.1111/j.1365-2648.2006.03809.x

Teasdale, J. D. (2004). Mindfulness-based cognitive therapy for depression: Replication and exploration of differential relapse prevention effects. *Journal of Consulting and Clinical Psychology*, 72(1), 31-40. doi: 10.1037/022-006X.72.1.31

Teasdale, J. D., Segal, Z. V., & Williams, J. M. (1995). How does cognitive therapy prevent depressive relapse and why should attentional control (mindfulness) training help? *Behaviour Research and Therapy*, 33, 25-39.

Telles, S. & Naveen, K. (2004). Changes in middle latency auditory evoked potentials during meditation. *Psychology and Reproduction*, 94, 395-400.

Teychenne, M., Ball, K., Salmon, J. (2008). Physical activity and likelihood of depression in adults: A review. *Preventative Medicine*, 46(5), 397-411; doi: 10.1016/j.ypmed2008. 01.009

Thompson, M., & Gauntlett-Gilbert. (2008). Mindfulness with children and adolescents: Effective clinical application. *Clinical Child Psychology and Psychiatry*, 13, 395-408. doi: 10.1177/1359104508090603

Toneatto, T., & Ngyuen, L. (2007). Does mindfulness meditation improve anxiety and mood symptoms? A review of the controlled research. *Canadian Journal of Psychiatry*, 52(4), 260-266.

Trulson, M. E. (1996). Martial arts training: a novel 'cure' for juvenile delinquency. *Human Relations*, 39, 1131-1140.

Twemlow, S. W., & Sacco, F. C. (1998). The application of traditional martial arts practice and theory in the treatment of violent adolescents. *Adolescence*, 33, 505-518.

Twemlow, S. W., Sacco, F. C., & Fonagy, P. (2008). Embodying the mind: Movement as a container for destructive aggression. *American Journal of Psychotherapy*, 62(1), 1-33.

Van der Klink, J. J., Blonk, R. W., Schene, A. H., & Van Dijk, F. J. (2001). The benefits of interventions for work-related stress. *American Journal of Public Health*, 91, 270-276.

Vancampfort, D., Probst, M., Skjaerven, L. H., Catalan-Matamoros, D., Lundvik-Gyllensten, A., Gomez-Conesa, A.,...De Hert, M. (2012). Systemic review of the benefits of physical therapy within a multidisciplinary care approach for people with schizophrenia. *Physical Therapy*, 92(1), 11-23.

Vancampfort, D., Vansteelandt, K., Scheewe, T., Probst, M., De Herdt, A., & De Hert, M. (2012). Yoga in schizophrenia: A systemic review of randomised controlled trials. *Acta Psychiatrica Scandinavica*, 126(1), 12-20. doi: 10.1111/j.1600-0047.2012.01800

Varambally, S., Gangadhar, B. N., Thirthalli, J., Jagannathan, A., Kumar, S., Venkatasubramanian, G. G.,....Nagendra, H. R. (2012). Therapeutic efficacy of add-on yogasana intervention in stabilized outpatient schizophrenia: Randomized controlled comparison with exercise and waitlist. *Indian Journal of Psychiatry*, 54, 227-232. doi: 10.4103/0019-5545-102414

Veehof, M. M., Oskam, M. J., Schereurs, K. M., & Bohlmeijer, E. T. (2011). Acceptance-based interventions for the treatment of chronic pain: A systematic review and meta-analysis. *Pain*, 152, 533-542. doi: 10.1016/j.pain.2010.11.002

Visceglia, E., & Lewis, S. (2013). Yoga therapy as an adjunctive treatment for schizophrenia: A randomized, controlled pilot study. *The Journal of Alternative and Complementary Medicine*, 17, 601-607. doi: 10.1089/acm.2010.0075

Wall, R. B. (2005). Tai Chi and mindfulness-based stress reduction in a Boston public middle school. *Journal of Pediatric Health Care*, 19(4), 230-237. doi: 10.1016/J.pedhc.2005..02.006

Wang, C. (2011). Tai Chi and rheumatic diseases. *Rheumatic Diseases Clinics of North America*, 37(1), 19-32. doi: 10.1016/j.rdc,2010.11.002

Wang, C. (2012). Role of Tai Chi in the treatment of rheumatologic diseases. *Complementary and Alternative Medicine*, 15(6), 598-603. doi: 10.1007/s11926-012-0294-y

Wang, S. M., & Kain, Z. N. (2001). Auricular acupuncture: A potential treatment for anxiety. *Anesthesia and Analgesia*, 92, 548-553. doi: 10.1213/00000539-200102000-00049

Weber, S. (2004). The effects of relaxation exercises on anxiety levels in psychiatric inpatients. *Journal of Holistic Nursing*, 14, 196-205.

West, J., Otte, C., Geher, K., Johnson, J., & Mohr, D. C. (2004). Effects of Hatha yoga and African dance on perceived stress, affect, and salivary cortisol. *Annals of Behavioral Medicine*, 28(2), 114-118. doi: 10.1207/s15324796abm2802-6

White, A., & Moody, R. (2006). The effects of auricular acupuncture on smoking cessation may not depend on the point chosen—An exploratory meta-analysis. *Acupuncture in Medicine*, 24, 149-156.

White, L. S. (2012). Reducing stress in school-age girls through mindful yoga. *Journal of Pediatric Health Care*, 26(1), 45-56. doi: 10.1016/j.pedhc.2011.01.002

Whitehead, W. (2006). Hypnosis for irritable bowel syndrome: The empirical evidence of therapeutic effects. *International Journal of Clinical and Experimental Hypnosis*, 54, 7-20.

Wilk, C., & Turkoski, B. (2001). Progressive muscle relaxation in the cardiac rehabilitation: A pilot study. *Rehabilitation Nursing*, 26, 238-243.

Williams, K. A., Petronis, J., Smith, D., Goodrich, D., Wu, J., Ravi. N., ... Steinberg, L. (2005). Effect of Iyengar yoga therapy for chronic low back pain. *Pain*, 115(1-2), 107-117. doi: 10.1016/j.pain.2005.02.016

Williams, J. M., Russell, I., & Russell, D. (2008). Mindfulness-based cognitive therapy: Further issues in current evidence and future research. *Journal of Consulting and Clinical Psychology*, 76, 524-529.

Wolsko, P. M., Eisenberg, D. M., Davis, R. B., & Phillips, R. S. (2004). Use of mind-body medical therapies: Results of a national survey. *Journal of General Internal Medicine*, 19, 43-50. doi:10.1111/j.1525-1497.2004.21019.x

Wynd, C. A. (2005). Guided health imagery for smoking cessation and long-term abstinence. *Journal of Nursing Scholarship*, 37, 245-250.

Yadav, R. K., Magan, D., Mehta, N., Sharma, R., & Mahapatra, S. C. (2012). Efficacy of a short-term yoga-based lifestyle intervention in reducing stress and inflammation: preliminary results. *Journal of Alternative and Complementary Medicine*, 18(7), 662-667. doi: 1089/acm.2011.0265

Yalom, I. D. (2005). *The theory and practice of group psychotherapy*. New York, NY: Basic Books.

Yeh, M., Lee, T., Chen, H., & Chao, T. (2006). The influences of Chan-Chuang Qi Gong therapy on complete blood cell counts in breast cancer patients treated with chemotherapy. *Cancer Nursing*, 29, 149-155.

Yook, K., Lee, S., Ryu, M., Kim, K., Choi, T. K.,....Suh, S. Y. (2008). Usefulness of mindfulness-based cognitive therapy for treating insomnia in patients with anxiety disorders: A pilot study. *Journal of Nervous and Mental Diseases*, 196, 501-503.